Reproductive Physiology and Aquaculture

THE EDITOR

Dr. Sunil Kumar P., Assistant Professor in the Department of Zoology at Sree Kerala Varma College, Thrissur, Kerala affiliated to the Calicut University. He has secured Ph.D in Fisheries from Central Institute of Fisheries Education, Mumbai a Deemed University under ICAR, New Delhi. He has published more than 25 research papers in referred journals and has few conference proceedings and book chapters to his credit. The book covers two sections; Reproductive physiology and Aquaculture. The first section incudes 11 chapters and second one covers chapters on live feed culture, zebrafish, antimicrobial resistance and egg yolk antibody.

Reproductive Physiology and Aquaculture

— Editor —

Sunil Kumar P.

— Authors —

K. J. Abraham
Binu Ramachandran
Reshma Silvester
Thushara James

2024

Daya Publishing House®
A Division of
Astral International Pvt. Ltd.
New Delhi – 110 002

ISBN: 978-93-5919-186-7

Published by : **Daya Publishing House®**
A Division of
Astral International Pvt. Ltd.
– ISO 9001:2015 Certified Company –
4736/23, Ansari Road, Darya Ganj
New Delhi-110 002
Ph. 011-43549197, 23278134
E-mail: info@astralint.com
Website: www.astralint.com

Preface

The first section on Reproductive Physiology deals with the different physiological processes involved in the reproduction and breeding of fishes. One of the major criteriain selecting species for aquaculture is the availability of suitable techniquesfor controlled or captive breeding. Controlled breeding is possible only with an adequate knowledge of the factors governing the reproduction of an animal andits breeding behaviour. Controlled breeding is also needed for the domesticationof the animal and for taking advantage of the benefits of genetic selection and hybridization.

The study of reproduction in fishes has also practical importance in Fish Biology and Stock Assessment, in solving some Fishery Management questions, such as the determination of Spawning Stock, Reproductive output and Recruitment. The topics under Reproductive Biology of Fishes, in the present book, ranges from basic information on Reproductive Biology *viz*; Reproductive organs in Teleosts and Crustaceans, Sexual dimorphism, Reproductive Strategies, Gametogenesis, Embryonic Development in Fishes, to, topics related to breeding of fishes *viz*, hormones involved in reproduction and Induced breeding. The Chapters on Neuroendocrine systems in Crustacea, give insights into Crustacean Reproductive Physiology, while, the chapter on Cryopreservation of gametes, outlines the basic principles and procedures involved in long term preservation of genetic material.

The chapter on live feed culture includes description on various live feed organisms microbes, microalgae, yeast based diets, newly hatched artemia, enriched artemia, rotifers, copepods *Proales similis*, Calanoid copepod, *Euterpina acutifrons*, *Parvocalanus crassirostris*, cladocerans, tubifex chronomid larva, nematodes, marine infusoria like *Fabrea salina* and Euplotes, microencapsulated diets and notes on constraints in live feed culture and their future prospects.

Zebrafish (*Danio rerio*, Hamilton, 1822) have gained enormous popularity as model organisms in biomedical research during the past few decades. As a promising model organism in developmental biology, genetics, physiology, and toxicity, this small aquatic vertebrate fish species has come to light. This fish has recently entered the field of neuroscience and biological psychiatry and has swiftly established itself as a key model animal. It is not unexpected that humans and fish have a great deal in common physically and behaviourally given that they share a high genetic similarity with humans (~75% based on coding regions). The fact that zebrafish can be a great model for human neuropsychiatric disorders should therefore not be surprising.

The current book chapter "Zebrafish: A promising model in research" discusses a wide range of zebrafish aspects that have fascinated the interest of scientists who are primarily interested in zebrafish behavioural research. The chapter begins by introducing zebrafish developmental stages before moving on to the care required for a balanced and healthy laboratory maintenance of fish population, an important section that all researchers are looking forward to. This chapter addresses the pertinent questions raised by the scientific community, such as why zebrafish? What is the current state of zebrafish research on a global and national scale? The chapter concludes with a list of the most common zebrafish behaviours observed in the laboratory as well as in nature. This section will assist the interested research community in comprehending the fundamental level behavioural parameters and their significance in their respective study area. The chapter includes images of a three rack automated fish facility for the efficient maintenance of zebrafish populations in the laboratory as well as the photographs of zebrafish embryos representing various developmental stages that were taken using Leica stereo microscope.

The authors guarantee that every reader will receive accurate, general, and valuable information on zebrafish. We cordially invite anyone who is interested to experience the scientific reading.

This chapter on antimicrobial resistance in global aquaculture deals with the global issue of antimicrobial resistance (AMR) in the aquaculture farms. The role of imprudent antibiotic usage and other factors in driving the emergence of AMR in these farms are been discussed. The various pathways through which AMR is transmitted in a typical aquaculture farm is detailed here. And finally, the various alternative approaches such as vaccines, probiotics, bacteriophage therapy etc. which can be effectively adopted to reduce the dependency on antibiotics and thereby combat AMR in aquaculture has also been suggested.

The last chapter deals with Ig Y antibodies. The aquaculture industry has a history of thousands of years. The volume of global fish production amounted to 174.6 million metric tons in 2020, up from 148.1 million metric tons in 2010. Infection control in livestock is a major concern for all farmers. Farms are always troubled by the existence of atypical microflora, pathogenic microflora, and multidrug-resistant microflora. In the instance of aquaculture, the seriously damaged fishes were discovered only after the terminal symptoms of swimming abnormalities and mass fish mortality. As a result, farmers are constantly required to use antibiotics to combat illness. The most commonly used antibiotics are synthetic chemicals in nature, and they leave residues in the environment with which they come into contact, either directly or indirectly. Researchers and farmers have long concentrated on using environment friendly methods to treat or control such illnesses. We'd like to give an idea about how IgY technology can be used in aquaculture. The IgY antibodies are simple to produce in large quantities at a low cost, and there is no need to harm or kill the birds; the percentage of pure antibody extraction is high; they are truly biological in nature, and do not cause environmental issues.

Sunil Kumar P.

Contents

SECTION 2: AQUACULTURE

Section 1

Reproductive Physiology

Chapter 1

Reproductive Organs in Fishes, Crabs and Prawns

Introduction

Reproductive Physiology deals with the different physiological processes involved in the reproduction and breeding of fishes. One of the major criteria in selecting species for aquaculture is the availability of suitable techniques for controlled or captive breeding. Controlled breeding is possible only with an adequate knowledge of the factors governing the reproduction of an animal and its breeding behaviour.

Controlled breeding is also needed for the domestication of the animal and for taking advantage of the benefits of genetic selection and hybridization.

Reproductive System of Crab

Crabs are dioecious or unisexual (occurrence of separate males and females), and the sex can be identified by a number of external and internal characters. The males generally grow to a larger size and their chelate legs are bigger, compared to females of the same size.

1. Male Reproductive Organs

The reproductive organ system of male crab is bilaterally symmetrical, creamy to whitish in colour and composed of a pair of **testes**, a pair of **vasa deferentia** and a pair of **ejaculatory ducts**, internally. Externally, the reproductive system consists of a pair of **penis** and a pair of primary and secondary pleopods, as accessory reproductive organs, present on the inner side of the abdominal flap. The first and second abdominal appendages are highly modified to function as copulatory organs.

A. Testis

The testis is an opaque organ with an irregular surface. The testes are tubular and highly coiled, paired organs, roughly in the form of the English alphabet **'H'** located on the dorsal part of the hepatopancreas, sandwiched between the hepatopancreas and hypodermis of the carapace. It continues laterally up to the level of the vas deferens. Just before the posterior stomach and anterior to the heart, the left and right testes are connected together by a short commissure, a little ahead of the middle of the carapace.

Each testis has the appearance of a thick convoluted tube and is an opaque organ with an irregular surface. The testes are composed of numerous follicles, which are closely packed together. Each follicle has a thin wall enclosing a large cavity, which is continuous with that of the vas deferens. Each follicle or testicular lobule is surrounded by an outer layer of thin connective tissue and an inner layer of epithelium. The lumen of the tubule contains generative cells and their progeny (spermatogonia, spermatocytes, spermatids and sperms)

The lobules open into the seminiferous duct, which gradually becomes a short muscular structure called **vasa efferentia**, delivering spermatozoa into the anterior vas deferens.

B. Vas Deferens

The vas deferens arises from the posterior end of the testis and occupies a large part of central body cavity. The vasa deferentia include a pair of elongated, convoluted tubes, passing backwards from the posterior part of testis to the posterior part of the thorax. Each vas deferens opens to the exterior, on the coxa of the last peraeopod or walking leg.

The vas deferens is divisible into 3 distinct regions

1. Anterior Vas Deferens (AVD)

It is the part of vas deferens which arises from the posterior end of the test is near the commissure, on either side of the median line of the cephalothorax.

It is a narrow tube like structure, tightly coiled and bound together as a single mass. The coils are so fused together that it cannot be straightened or separated. It rests on the massive median vas deferens. The anterior vas deferens is opaque and whitish in colour, due to the presence of sperms. The coils of Anterior Vas Deferens increase in thickness posteriorly and lead into the Median vas deferens.

The spermatozoa accumulated in the seminiferous duct of the testes are transferred to the AVD, where they are packed and stored as spermatophores by the secretions of the AVD. The spermatophores are ejaculated outside through the ejaculatory duct.

2. *Median Vas Deferens (MVD)*

is the biggest portion of the vas deferens. It consists of bigger coils which are fused to each other and difficult to separate. It is also filled with a white viscous fluid.

3. *Posterior vas deferens (PVD)*

The Median Vas Deferens leads into a thin walled and transparent posterior vas deferens. It is a long and convoluted tube, of lesser diameter than the Median Vas Deferens, but with a higher diameter than the AVD. PVD is wider at the anterior region, but narrows in the posterior region to form the ejaculatory duct. It is filled with a clear fluid. The lumen of the PVD also contains spermatophores, which are ovoid in shape. In the posterior region, it narrows to form an ejaculatory duct.

C. Ejaculatory Duct

The ejaculatory duct is a relatively straight, narrow and smooth transparent tube, which passes through the musculature of the coxa of the fifth walking leg and opens at the base of the coxal segment, through the penis.

D. Penis

Penis is the terminal portion of the ejaculatory duct. It arises from the ventro-median part of the coxa of the 5th peraeopod. It is a weak and slender tube, with a gonoduct at its centre. It lies between the ventral surface of the thorax and the abdomen and is found inserted in the foramen of the first pleopods.

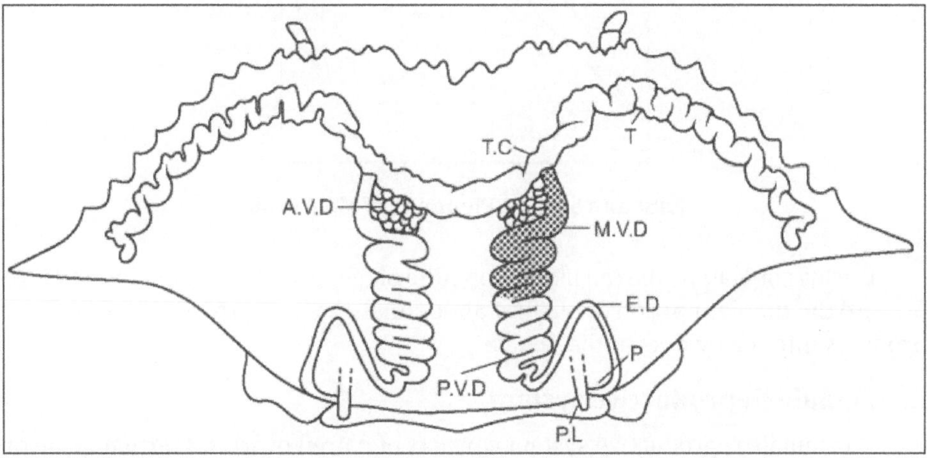

AVD: Anterior vas deferens; MVD: Median Vas Deferens; TC: Transverse commissure; PVD: Posterior Vas Deferens; ED: Ejaculatory duct; P: Penis; PL: First pleopods

Male Reproductive System.

Accessory Structures of Reproduction/Secondary Sex Organs

In males there are only two pairs of abdominal appendages, which are situated on the first and second abdominal segments and they act as copulatory organs

The highly specialised first pleopod of the male is the functional intromittent organ. The first abdominal appendage has two joints; the basal part called protopodite is broad and flattened on the inner side, the distal elongated portion is the endopodite, which is long and tapering with a curved outline. The first pleopod is a deeply grooved structure in which the edges of the groove are fused to form a tube. On the inner surface of the base of the endopodite is a trough like area, which receives the spermatophores and seminal fluid from the penis and the closed groove acts as a tube during copulation to transfer these materials into the seminal receptacles through the oviduct of the female.

The second abdominal appendage also consists of two parts; the basal portion is the protopodite. The endopodite is a vertical rod like structure attached to the protopodite.

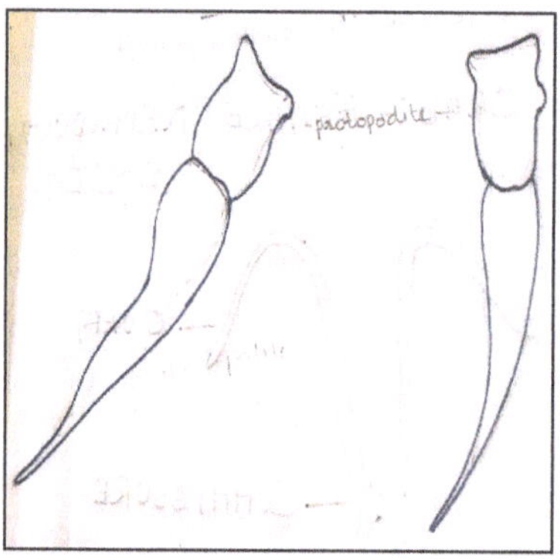

First and Second Pleopods of Male Crab.

During copulation, the rod like endopodite of the second abdominal appendage fits into the tubular portion of the first abdominal appendage to force the seminal products into the oviduct of the female.

2. Female Reproductive system

The female reproductive system consists of paired ovaries, spermathecae or seminal receptacles and paired oviducts.

Ovaries

Ovaries are paired structures occupying similar position as the testis of the male *ie*, beneath the hypodermis and above the hepatopancreas. Each ovary may be divided into lobes and the right and left lobes are unequal in most cases or may be fused together. The anterior lobes of the ovary extend up to the base of the anterio-lateral spines and posteriorly, it extends up to the posterior end of the thorax.

The paired ovaries are connected posteriorly behind the pyloric foregut by a commissure which forms a 'H' shaped structure. Posterior to the commissure, the ovary gives rise to a large sac, called spermathecae or **seminal receptacle.**

The spermathecae arise from the lateral border of the posterior lobe of the ovary and is actually an enlarged portion of the oviduct. On the ventral side, the spermatheca is continued into a short tube called **oviduct**.

The oviduct passes through the muscles of the third peraeopod and opens to the exterior on the sternum of the 6th thoracic segment. The spermatheca acts as a storage organ for the sperms after copulation.

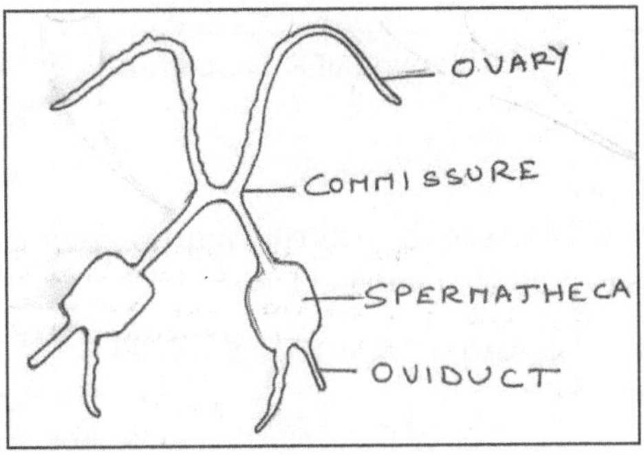

Crab – Female Reproductive System.

Accessory Structures

In females there are four pairs of abdominal appendages or pleopods that function as accessory reproductive structures. They are located on the second to fifth abdominal segments.

Each appendage has a biramous structure, consisting of a basal protopodite from which arise the exopodite and endopodite. The second to fourth appendages present on the third to fifth abdominal segments are similar in structure. The exopodites are modified into an elongated fan like structure having a large number of setae on the anterior and posterior borders. The endopodite is segmented and bears clusters of setae on each segment. The ova are attached to the long setae of the endopodites after ovulation and spawning.

In the first pair of pleopods, the exopodites are cylindrical and has rows of setae. The endopodites are as long as the exopodites and have transverse grooves. Many long setae arise from these transverse grooves.

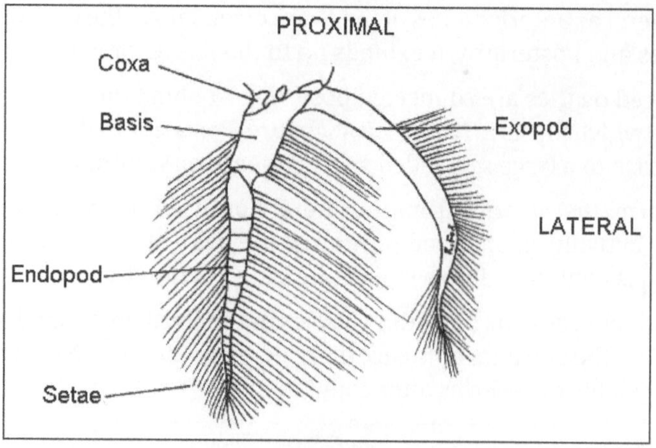

4th Pleopod of Female Crab.

Sexual Dimorphism

Sexual dimorphism refers to the external morphological characters by which the sexes of a species can be distinguished.

Sexual Dimorphism in Crab

Character	*Male*	*Female*
Shape of the abdomen	Inverted 'T' shaped	Broadly triangular in immature crabs; becomes semicircular in adults
Abdominal Appendages	Two pairs of abdominal appendages, present on the first and second abdominal segments, needle like in structure and modified as copulatory organs	4 pairs of abdominal appendages present on the 2nd to 5th abdominal segments, carrying setae for holding eggs.
Position of Genital opening	Genital opening on the coxa of the 5th peraeopod	Genital opening on the sternum of the 6th thoracic somite between the 3rd pair of walking legs.

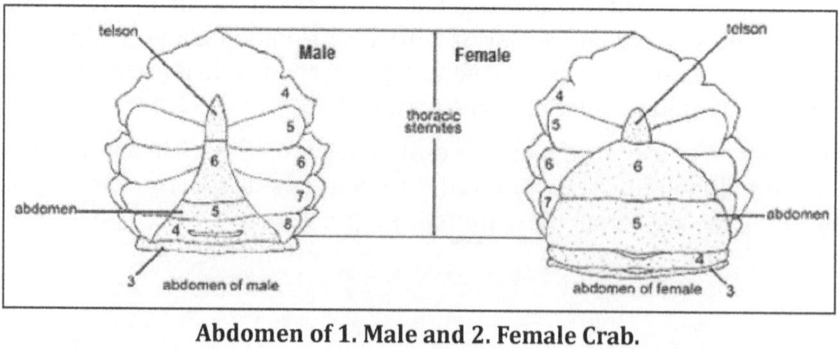

Abdomen of 1. Male and 2. Female Crab.

Mating and Fertilisation

Mating occurs between a post moult female and a hard male. The pair of crabs engage in a pre- mating embrace, with the female on its back. The male crab wraps its claws and legs around the female. The abdominal flaps of both partners will be flexed open. The embrace last for several days before the actual mating takes place. A pair of mating crabs is known as a 'doubler'.

The male transfers the sperms by inserting its abdominal appendages into the two genital pores on the underside of the female. The female stores these sperms in a storage sac called spermatheca, until she is ready to use it.

After copulation, the cavity of the oviduct seals by a secretion that hardens in contact with sea water. The sperms remain viable in the female's spermatheca for several weeks or months till the eggs mature. As the eggs pass through the spermathecal openings, in the paired oviducts, they are fertilized. The extruded eggs are attached to the setae of the pleopods and are carried by the female till they hatch.

Prawn – Reproductive System

Prawns are dioecious, the male and female prawns can be distinguished with the help of external organs.

Male Reproductive System

Male reproductive system consists of internal organs – paired testes, paired vas deferens, paired terminal ampoules and external organs – Petasma and a pair of Appendix Masculina.

Testes

The testes are paired, unpigmented and translucent organs. They lie in the thorax, one on either side of the middle line under the carapace or dorsal to the hepatopancreas. The testes are tubular and united in front. Each testis consists of 5 lateral lobes or caecal diverticula on each side. The lobes are connected to each other at their inner ends and leads to the next part, the Vas Deferens.

Vas Deferens

The Vas Deferens arises from the posterior margin of the testes. Each vas deferens consists of 3 regions

a. A short and narrow proximal vas deferens
b. A thick and large median vas deferens
c. A long and narrow distal vas deferens.

a. Proximal Vas Deferens (PVD)

It is a short and slender tube arising from the posterior margin of the testes. Its main function is to transport sperms from the testes to the median vas deferens by peristaltic movements. The PVD has an outer connective tissue coat and thick circular muscle fibres lining the inner wall of the duct. No glandular cells are present in PVD. The PVD dilates into a blind pouch, of the Median Vas deferens, which in turn continue anteriorly as the ascending limb of the MVD.

b. Median Vas Deferens (MVD)

Consists of an ascending limb and a descending limb. The ascending limb of Median Vas Deferens is divided internally into 2 unequal ducts by a connective tissue septum; the larger **Sperm Duct** and a smaller **Wing Duct.** The sperm mass is packed into **Spermatophores** by the glandular secretions of the median Vas Deferens. The wing duct secretes the materials for the spermatophore wing. The blind pouch of the MVD functions for the storage of sperms.

c. Distal Vas Deferens (DVD)

It is a long and slender tube which leads into the **Terminal Ampoule** and functions to convey the spermatophores from the Median Vas Deferens to the terminal ampoule.

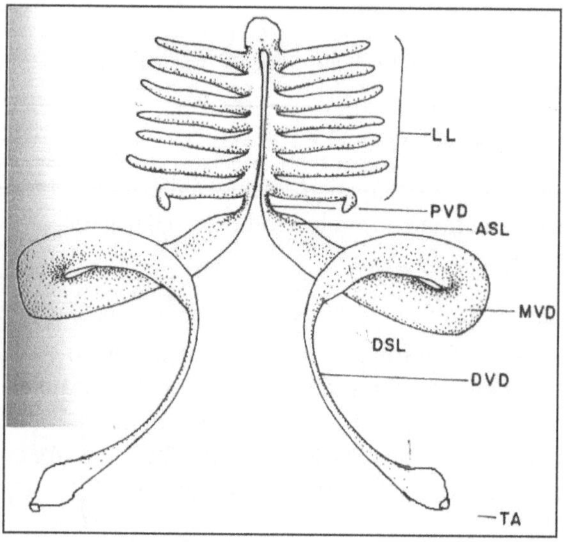

LL: Lateral lobes; PVD: Proximal vas deferens; ASL: Ascending limb; DSL: Descending limb; MVD: Median Vas Deferens; DSL: Distal vas deferens; TA: Terminal ampoule.

Prawn – Male Reproductive System.

Terminal Ampoule

It is a bulbous structure with thick muscular walls embedded in the muscles of the coxa of the 5th peraeopod. The walls are lined with secretory epithelial cells. It appears opaque and white due to the presence of fully developed spermatophores inside.

The final and complete moulting of the spermatophores takes place inside the Terminal ampoules by its secretions. The muscular walls of the terminal ampoules help in the forceful ejaculation of the spermatophores.

Structure of the Spermatophore

A spermatophore is roughly oval in shape. It consists of two parts- a **body** and a **wing.** Externally the body of the spermatophore has a dirty cream colour and is thrown into folds, resembling a bag full of grapes. The body has a translucent parachute like wing, connected to the body by means of a short stalk.

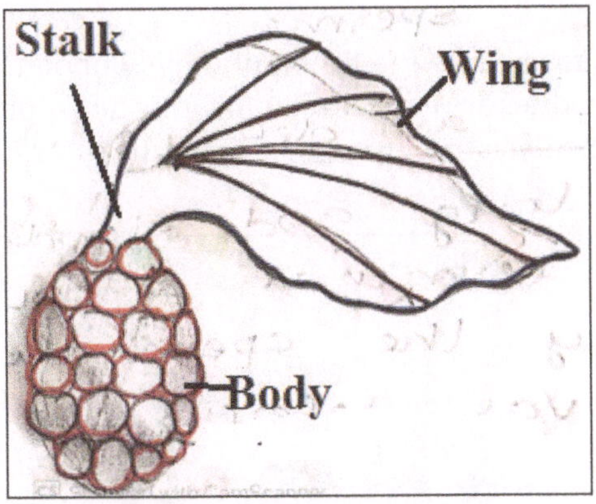

Spermatophore of Prawn.

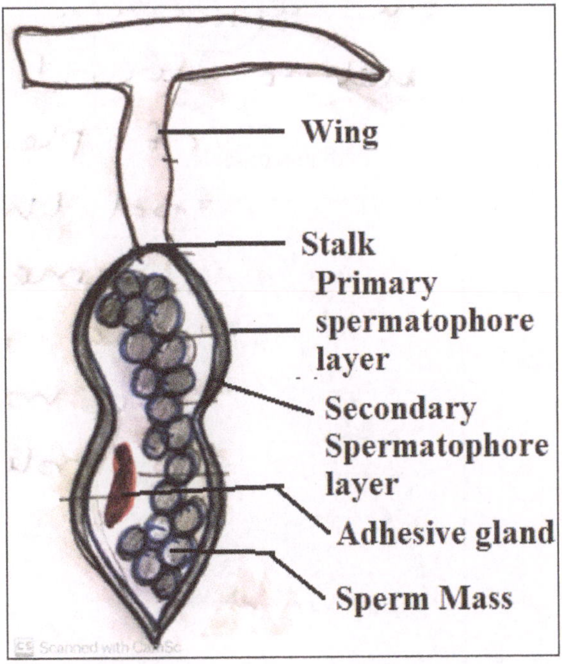

Longitudinal Section of a Spermatophore.

The sperm mass is embedded within a matrix in distinct oval lobules. The spermatophore is enclosed in two layers of secretions produced by the gland cells of the vas deferens. An adhesive gland is present on one side of the sperm mass, which helps in sticking the spermatophore in the thelycum. The wing helps in anchoring the spermatophore inside the thelycum of the female, after which it dissolves in the surrounding sea water.

Accessory Reproductive Structures in Male

Males are distinguished externally by the presence of the petasma and appendix masculine.

The Petasma is formed by the fusion of the endopodites of the first pair of pleopods. The endopodites are membranous bearing hooks on the inner side which interlock to form a median structure. It serves for the transfer of spermatophores during copulation.

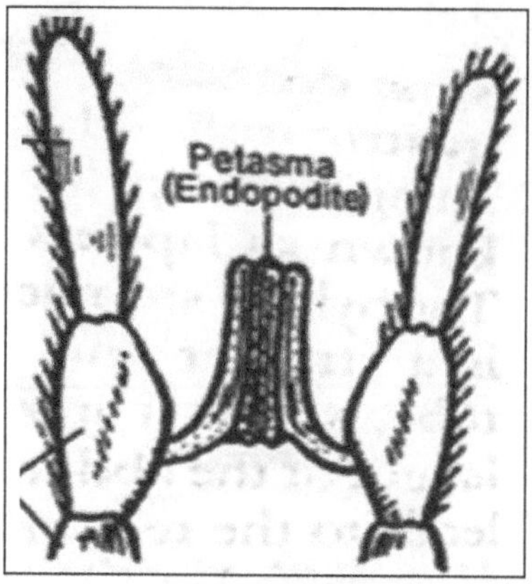

Petasma of Male.

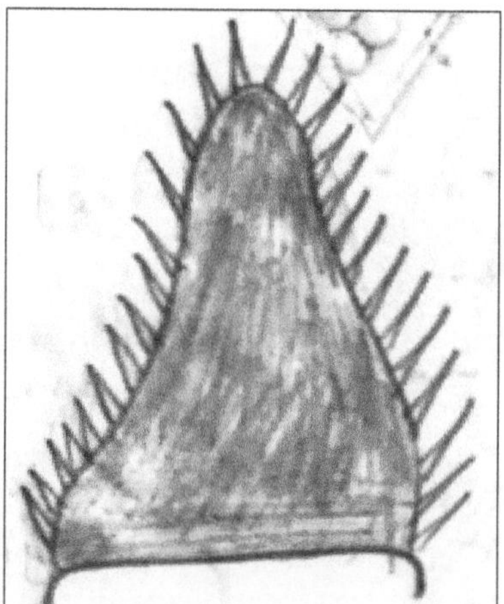

Appendix Masculine.

The **Appendix Masculina** is a small structure formed on the endopod of the second pair of pleopods. It is almost oval in shape and the outer margin is fringed with numerous spines. The appendix masculine is believed to assist the petasma in the transference of spermatophores into the female.

Female Reproductive System

The female reproductive system consists of paired internal **ovaries, oviducts** and the external **Thelycum.**

There are two bilaterally symmetrical ovaries, one on either side of the middle line. The mature ovary extends the entire length of the body from the base of the rostrum anteriorly to the base of the telson posteriorly. The right and left ovaries are united posteriorly and free anteriorly.

Three main lobes are present in each half of the ovary. The **anterior lobe** occupies the cephalic region, the **middle lobe** has 5-7 finger like **lateral lobules** or **diverticula**, on each side and fills the thoracic region above the hepatopancreas. The **posterior lobes** are united along their entire length and extend up to the telson. Each half of the anterior and middle lobes are united only in two regions; at the base of the middle lobes and at the base of the anterior lobes.

The oviducts are a pair of short and narrow tubes arising from the penultimate lobule of the middle lobe of the ovary and opens out by gonopores or genital apertures on the coxa of the third pair of peraeopods or walking legs.

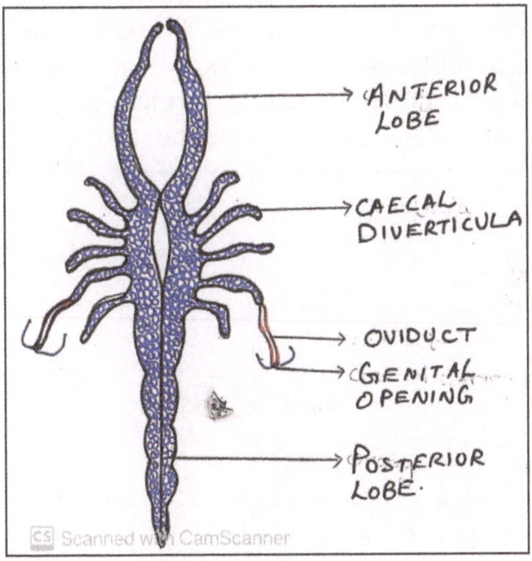

Prawn – Female Reproductive System.

Accessory Reproductive Structures in Female

The accessory reproductive organs of the female prawn are the **thelycum**, which act as the seminal receptacle of the female. The thelycum consists of modified sternal plates of the 7th and 8th thoracic somites. It includes an anterior plate and a pair of lateral plates. The lateral plates meet each other posteriorly in the median line and the edges of the plate form an elevated ridge. A 'V' shaped slit is formed between the lateral and median plates, into which the spermatophores are inserted during mating.

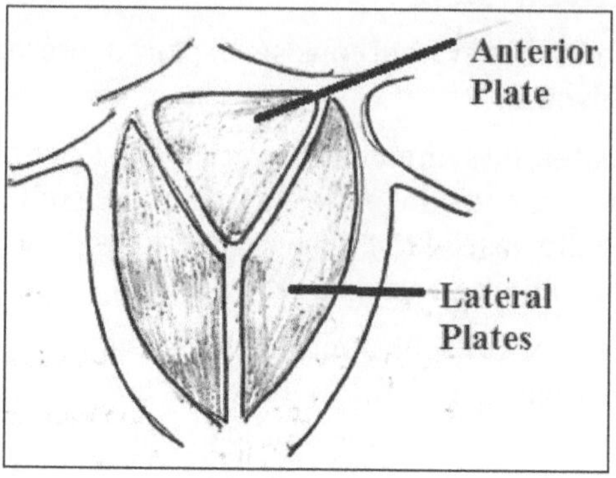

Thelycum

Reproductive System of Fish

1. Male Reproductive System

In male fishes, reproductive organs consist of a pair of testes situated on either side of the body cavity, ventral to the kidneys. The t**estes** are elongated, flattened structures attached to the body wall by **mesorchia**. From the posterior side of each testes, a pair of **sperm ducts** or **vasa deferentia** arises and runs posteriorly to open into the urogenital aperture, between the rectum and urinary bladder.

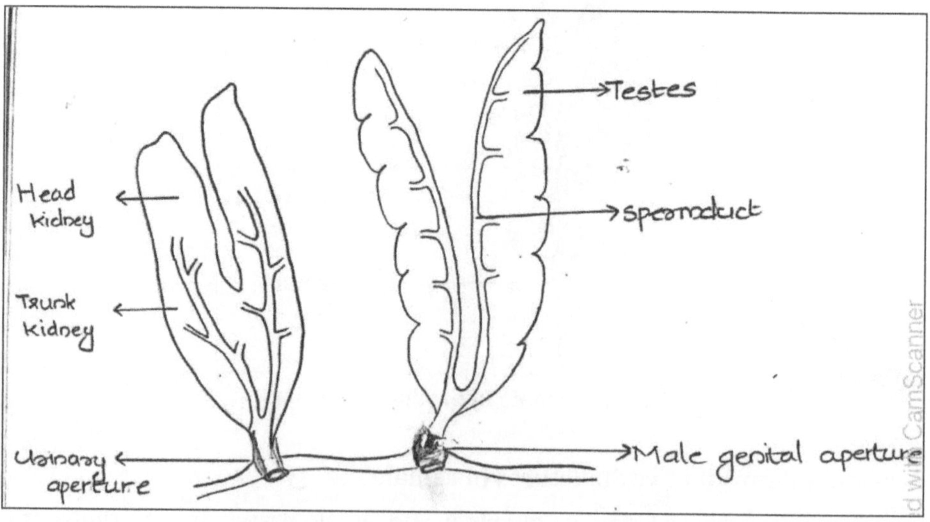

Male Reproductive System - Teleost.

In some teleosts, *e.g.*: *Clarius* and *Heteropneustes*, a pair of glandular structures called **seminal vesicles** are seen as outgrowths of the vasa deferentia. The seminal vesicles are secretory in nature and produce a fluid which nourishes the sperms and keep them in an active and viable condition.

Histology

The testes consist of a large number of lobules which contain highly convoluted **seminiferous tubules.** The seminiferous tubules are separated from each other by a stroma containing connective tissue cells, blood capillaries and interstitial cells. The seminiferous tubules open into a spermatic duct lined with secretory epithelium.

Inside the lobules of the testes, sperms are produced from **sperm mother cells** or **spermatogonia**.

Interstitial cells, corresponding to the Leydig cells of mammals are reported to be present in teleosts, producing male sex hormones.

The Lobule Boundary cells found in the walls of the lobules are believed to be homologous with the Sertoli cells of mammals and have a nutritive function or have a phagocytic action on the degenerating sperms.

2. Female Reproductive System

The reproductive organs of the female constitute the ovaries. In fishes ovaries are of two types.

1. Cystovarian ovaries.
2. Gymnovarian ovaries.

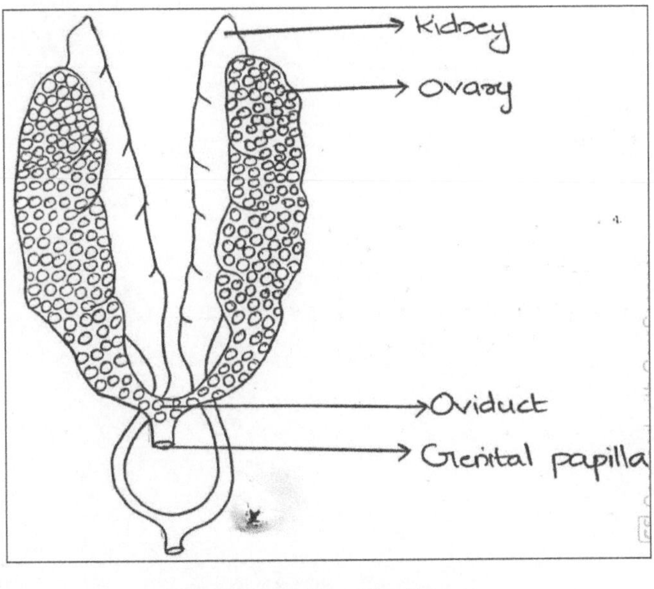

Female Reproductive System.

In Gymnovarian ovaries, the ova are discharged into the coelomic cavity, from where they pass out through the oviducal funnel. *e.g.*: Chondrichthyes and Dipnoi

In Cystovarian ovaries, the ovary is enclosed in a coelomic sac that is continued as an oviduct which opens to the exterior.

The ovaries are elongated, sac like, paired, hollow organs lying in the abdominal cavity, ventral to the kidneys and are attached to the body wall by **Mesovarium.**

The ovaries may be equal or unequal in size and each ovary is continued posteriorly into a short oviduct. The two oviducts join and open to the exterior by a common urinogenital opening.

Histologically, the wall of the ovary is composed of 3 layers.

1. **Outermost peritoneum**
2. **Tunica albuginea:** Made up of connective tissue, muscle fibres and blood capillaries.
3. **Germinal Epithelium:** This project into the cavity of the ovary (ovocoel) in the form of lamellae. These ovigerous lamellae are the seat for the development of oocytes, which are visible in various stages of development. The germ cells or oogonia are found in clusters in the lamellae and originate from the germinal epithelium. An oogonium has a large nucleus and a thin layer of ooplasm. Each oogonium passes through a number of maturation stages to become a ripe ovum. Oocytes in various stages of development can be seen on the lamellae. The oocytes finally mature to form ripe eggs or ova.

Chapter 2

Reproductive Strategies in Fishes

According to reproductive habits, fishes can be classified into:

1. Egg layers (Oviparous)
2. Live bearers (Ovoviviparous and Viviparous)

Oviparity

I. Oviparity in Teleosts

Eggs of oviparous bony fishes are typically 1.5 – 3 mm in diameter, though some species of trouts and salmons have eggs exceeding 5mm. In catfishes, eggs may vary in diameter from 15- 25 mm. An egg cell consists of living protoplasm, a nucleus and variable quantities of yolk contained within a vitelline membrane.

According to the nature of buoyancy, 3 types of eggs can be distinguished:

1. Pelagic Eggs

They are also known as planktonic or floating eggs and are characterized by their lightness and buoyancy, smaller size and transparency. A conspicuous feature is the presence of a single large oil globule. Pelagic eggs are non adhesive and free.

2. Demersal Eggs

Demersal eggs are characterized by their large size and opacity, they contain more yolk, their outer egg membrane is adhesive, so that the eggs readily adhere to each other, or are attached to submerged objects.

3. Semi-buoyant Eggs

These types of eggs neither float nor sink, but remain suspended in the water column.

II. Oviparity in Elasmobranchs

Fertilisation is internal in elasmobranchs. Elasmobranch eggs are enclosed in a stout and horny egg shell secreted by the shell gland of the oviduct. The egg capsule is equipped with tendrils to attach it with a weed or stone. The egg cases of elasmobranchs are known as mermaid's purses. It is quadrangular in shape with the four corners drawn out into long and spiral contractile tendrils for attachment.

Ovoviviparity

In ovoviviparity the eggs are retained within the body of the females for subsequent development. Eggs develop inside the reproductive tract of the female- the oviduct in Elasmobranchs and *the* ovary in teleosts. During the gestation period, the eggs are protected from predation and live in physiologically regulated surroundings.

1. Ovo-viviparity in Elasmobranchs

In ovo-viviparous species, there are two reproductive strategies;

1. The eggs develop into young ones, depending only on endogenous yolk reserves; the mother provides only protection and water.

   e.g. *Squalus acanthias* (Spiny dogfish)

   Centrophorus sp.

2. The developing young ones take nutrients from the uterus without forming a connection between the uterine and embryonic tissues

   e.g. In Torpedo, the electric ray, the uterine mucosa becomes modified to produce a secretion called Embryotrophe or Uterine milk, which is an important source of food. The uterine secretions are directly taken into the digestive tract through the mouth and spiracles.

   In certain elasmobranchs, the lining of the uterus is glandular and thrown into very long filamentous processes, known as **uterine villi** or **Trophonemata.** These are richly vascular and secrete a nutritive fluid in **Myliobatidae** (Eagle rays) and **Trygonidae** (Sting rays). The fluid is rich in fats and is passed into the alimentary canal of the embryo through its spiracles.

   In butterfly rays (*Pteroplatea*), the highly glandular **trophonemata** or **uterine villi** are long and dip through the spiracles of the embryo and extend into the oesophagus to release its secretions directly into the gut.

   Other examples of ovo-viviparity in elasmobranchs are seen in the mackerel shark, *Lamna nasus* where the developing young eat the other eggs in the uterus (oophagy) or consumes other embryos (adelphophagy) as in the sand tiger shark (*Carcharhinus taurus*) and short fin mako shark (*Isurus oxyrinchus*).

2. Ovo-viviparity in Teleosts

The guppy, *Lebistes reticulates* is the best example for an ovo-viviparous species. The eggs are fertilized within their follicles, where they develop for some time. On rupture of the follicles the embryos are released into the ovarian cavity, where they complete their development, sustained by the yolk.

Another example for a completely ovo-viviparous fish is *Sebastodes paucispinis* (Stone fishes), which undergoes an ovoviviparous development within the ovarian cavity. Fertilisation occurs immediately after ovulation and embryonic development takes place within the ovarian follicles. They depend on the egg yolk entirely for development and hatch out just prior to spawning, when all yolk has been absorbed and they are 4-5 mm long.

Viviparity

In viviparity, the embryo develops inside the body of the mother and an intimate connection develops between the maternal and embryonic circulation for exchange of nutrients and waste products

1. Viviparity in Elasmobranchs

The placental or viviparous elasmobranchs belong to the families *Carcharhinidae* and *Sphyrnidae*.

In *Scoliodon* and *Squalus*, the eggs develop inside the uterus. The mucous lining of the uterus forms protective fluid filled compartments, one for each embryo. Initially, the gut of each embryo is connected by a tubular yolk stalk with a yolk sac, containing yolk for nourishment. When the yolk is exhausted, the yolk sac becomes shriveled and embedded in the uterine wall forming a **yolk sac placenta.** Later on blood vessels develop and the yolk stalk is replaced with a placental chord, which gives off numerous slender tubular outgrowths, the **appendicula.**

2. Viviparity in Teleosts

In viviparous teleosts, the young ones are nourished through some kind of placental connection with the mother. In teleosts, gestation occurs in the ovary, either within the follicle or in the ovarian cavity.

a. Follicular Gestation

In *Heterandria Formosa*, the embryos having a small supply of yolk are retained within the ovarian follicles. The walls of each follicle develop a bed of blood capillaries, which forms small processes called **villi.** These villi make intimate contact the external surface of the embryo, where the maternal and embryonic blood system forms a placenta through which the embryo is nourished.

b. *Ovarian Gestation*

In the fishes of the families **Jenynsiidae** and **Goodeidae**, the embryos carry a small amount of yolk, when discharged from the follicles into the ovarian cavity. In the ovarian cavity, the embryos undergo further development. The embryos of Jenynsiidae are initially sustained by the secretions of the ovary. Later the lining of the ovary makes forms highly vascular folds which make contact with the gills of the young, thereby forming a branchial placenta.

In some fishes of the family Goodeidae, vascular absorbing processes arise from the hind gut of the embryo, **The Rectal Trophotaenia**, which are bathed in the ovarian secretions from which nutrition is drawn.

Chapter 3

Sexual Dimorphism in Fishes

Sexual dimorphism refers to the characters and structures by which the males and females of a species can be distinguished.

The sexual dimorphic characters that enable the identification of sexes can be classified as **primary** and **secondary.** Primary sexual characters are those that are actually concerned with the reproductive process and production of gametes *ie* testes and its ducts in males and ovaries and oviducts in females.

Secondary sexual characters consist of the accessory organs of reproduction. They can be divided into:

1. Structures which facilitate the fertilization of ova
2. Structures for courtship or for fighting with other males

1. Structures Related to Fertilization

Copulatory Organs in Males

Many fish species possess copulatory organs for transferring sperms into the body of the females

In Chondrichthyes, males possess claspers to facilitate internal fertilization. It is formed by the modification of the first ray of the pelvic fin. It is a rod like structure having a groove or canal along its whole length. During copulation, the grooves form a canal for the spermatic fluid. Claspers are also known as **Myxopterygia.**

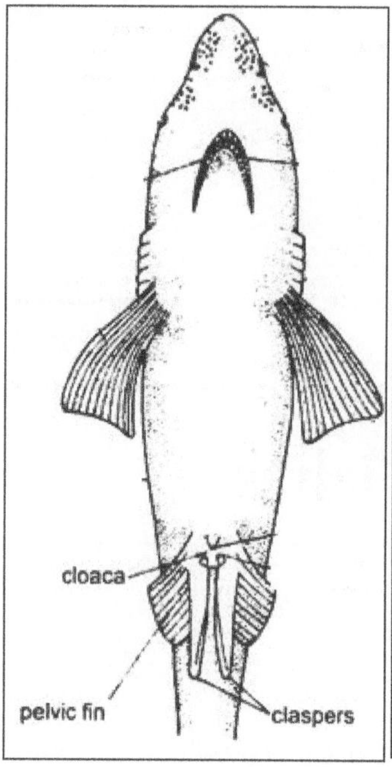

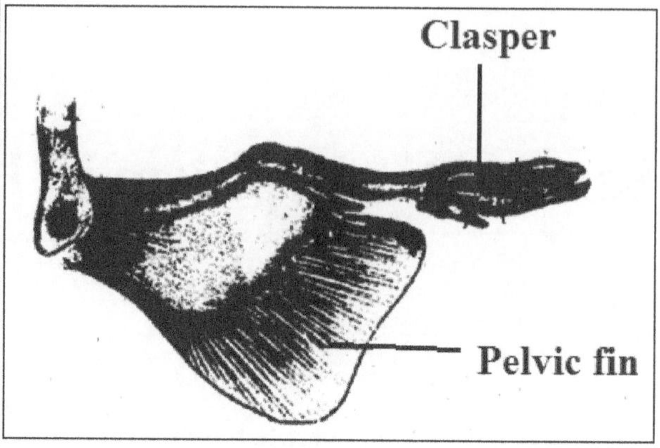

Claspers in Elasmobranchs. Clasper of Elasmobranch – Enlarged View of Pelvic Fin.

In Holocephali, *Chimaera*, a club shaped frontal clasper is present on the head of the male. The male is believed to use its frontal claspers to retain its hold on the females during copulation.

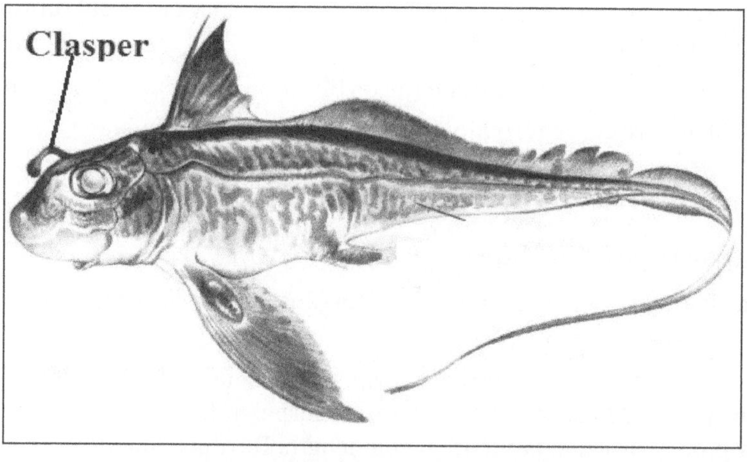

Club Shaped Frontal Clasper in Chimaera.

The anal fins become enlarged into an intromittent organ, the **gonopodium**, in mosquito fish (*Gambusia affinis*) and the Guppy (*Lebistes reticulates*).In mosquito fish, the vas deferens enters as a tube upto the anterior ray of the anal fin, which is modified to form an elongated tubular copulatory organ. In Poeciliidae (*e.g.* Guppy), the intromittent organ of anal fin is provided with hooks and barbs.

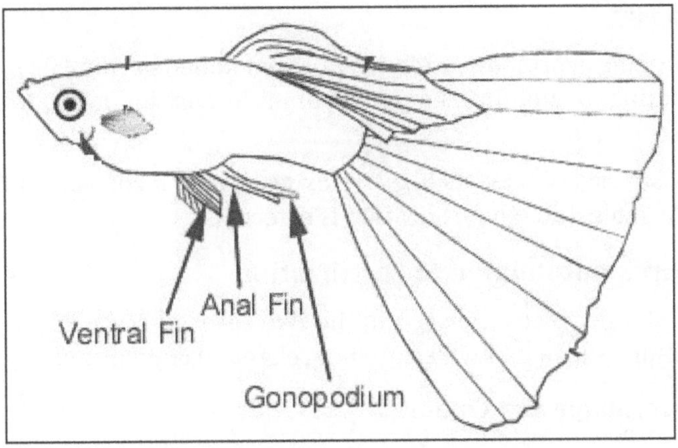

Gonopodium in Male Guppy.

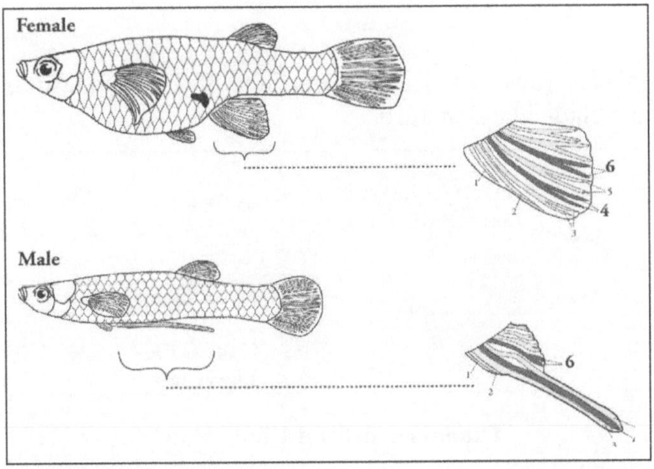

Male and Female Mosquito Fish (*Gambusia*) showing Gonopodium.

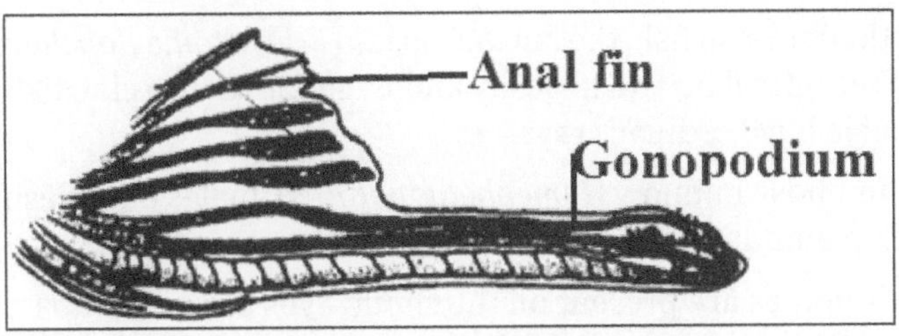

Gonopodium of *Gambusia* – Enlarged View of Anal Fin.

In *Catostomus* (white sucker), the anal fin is elongated to transfer milt into the female's body.

A penis like intromittent organ is present in the males of some marine Ophididae family.

In the clasping cyprinodont, *Xenodexia*, the modified pectoral fin or **Heterochir** is used in mating for holding the gonopodium in position for insertion into the oviduct of the female.

In a few species of teleosts like *Mystus seenghala*, a conical genital papilla is present in the male, though fertilization is external.

2. Structures not Related to Fertilization

Structural differences are seen in the two sexes of some species, especially during breeding season and which are not related to copulation.

a. Body Size, Shape and Colour

In most teleosts, the female is larger in size than males. Generally the female has a much more rounded belly than the male, especially when she is ripe for spawning. Males have a brighter colour of the body and fins during the breeding season.

b. Fins

In *Xiphophorus* (Sword tail), the lower lobe of the caudal fin is enlarged in males to form a blade like structure.

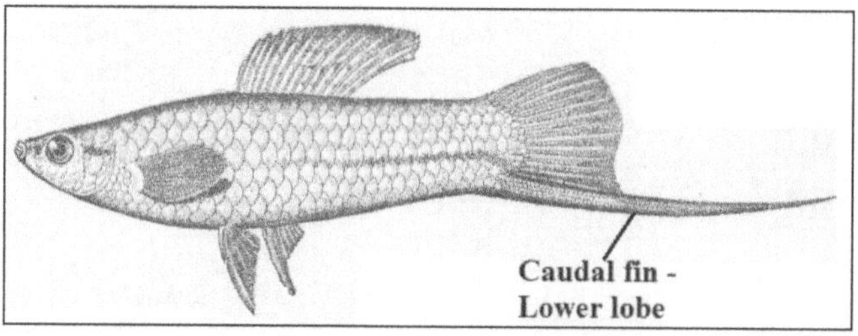

Caudal Fin of Sword Tail - Male.

The dorsal fin of males in *Molliensia* are enlarged to form a sail like structure, with coloured ocelli (eye spots).

In a North American fish, the fan tail darter (*Etheostoma flabellare*), the tips of the spines of the dorsal fin end in fleshy knobs, giving them a club like appearance, while the female have ordinary rays.

In the blunt nose minnow (*Pimephalus notatus*), males have fleshy rays, while those of the females are ordinary.

In *Bothus*, spines are present on the snout, eyes are wider apart and fin rays of the pectoral fin on the coloured side are elongated in males.

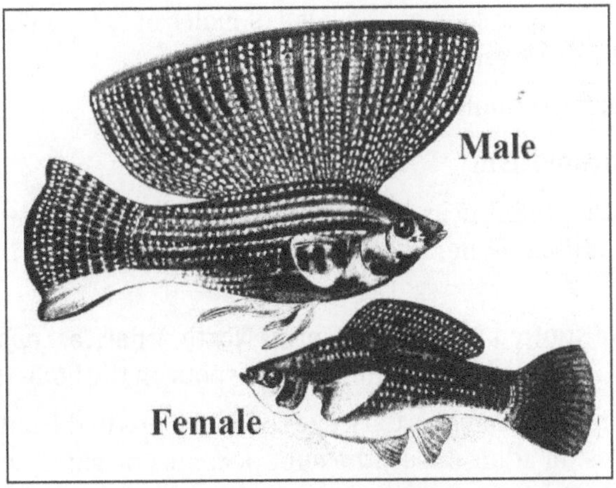

Enlarged Dorsal Fin in Males of *Molliensia* with Ocelli.

c. Tubercles

Pearl organs or nuptial tubercles appears in the males of the smelt (*Osmerus*), suckers and Cyprinidae during the breeding season and disappears shortly after breeding, under the influence of hormones.

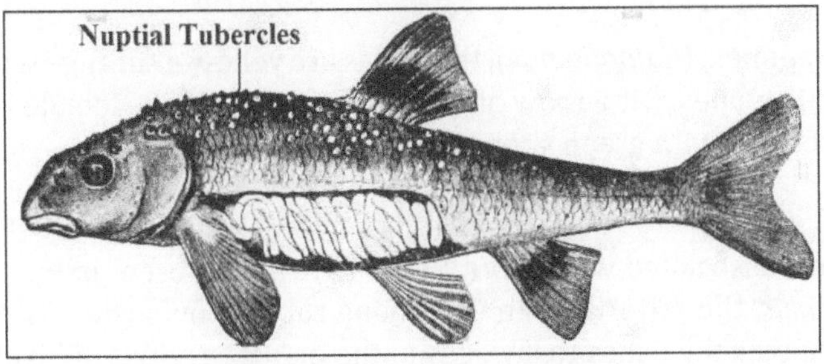

Nuptial Tubercles in Males.

d. Others

In male salmon, the lower jaw is enlarged and turned upwards, appearing like a hook during breeding season.

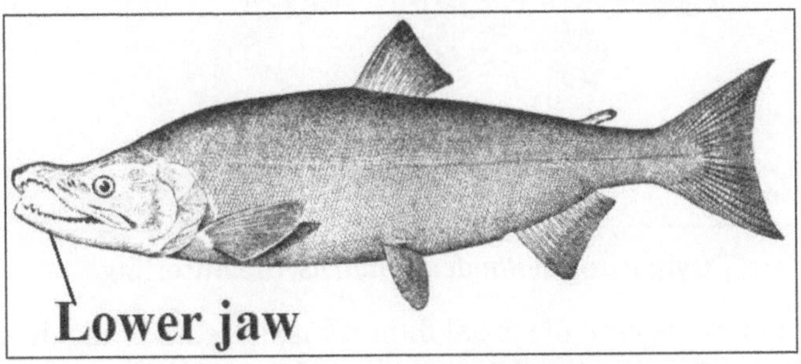

Enlarged Lower Jaw in Male Salmon.

A large fleshy lump is seen on the head of males of some cichlids, sparids (sea breams) and Labrids (Wrasses).

The males of most Gobies possess enlarged canines.

e. Sexual Dichromatism

It refers to the differences in colouration and colour pattern by which the sexes can be identified. Generally males are brighter or more intense in colour than females.

In the orange spotted sunfish of Central North America, *Lepomis humilis*, the male has more numerous and brighter orange spots on the body than the females.

In the bow fin (*Amia calva*), a dark eye spot develops in the tail region in young males and females; in adult males, the spot becomes brighter with an ocellation (eye like ring), while in females it becomes less prominent.

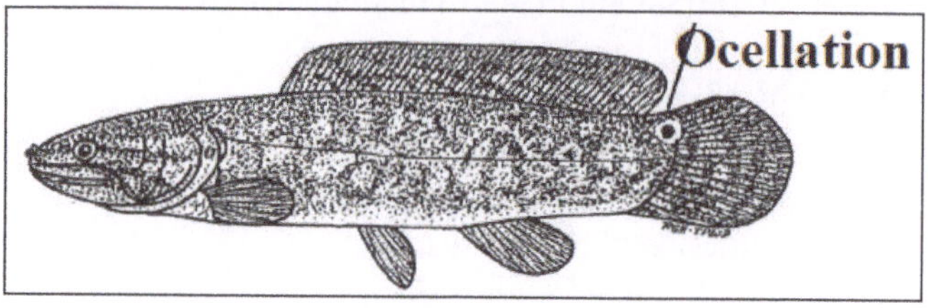

In the dragonets (*Callionymus*), the males are yellow or orange with deep blue stripes along the sides, with a row of blue or green spots. The female is dull yellow or brown in colour with green spots.

Accessory Reproductive Structures in Females

Structures associated with reproduction are also present in female fishes. In *Rhodeus amarus*, the bitterling, an egg laying tube or ovipositor is present. The ovipositor enables the fishes to lay eggs in the mantle cavities of bivalves.

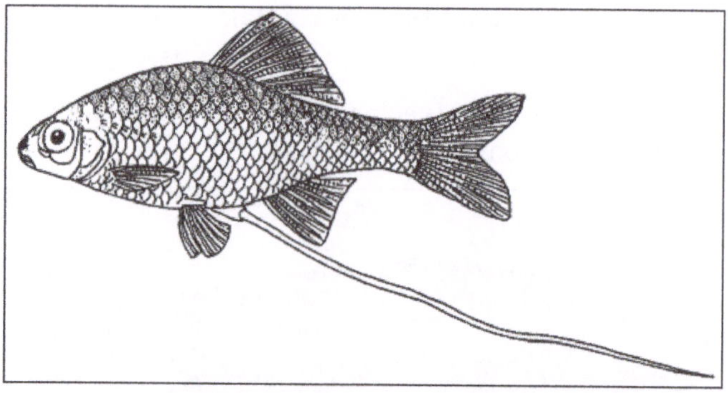

Ovipositor of *Rhodeus amarus*, the Bitterling.

The most extreme case of sexual dimorphism occurs in the deep sea angler fish, where the male is minute, all its organs are degenerate, except the gonads and it is parasitic on the female.

Chapter 4

Oogenesis

Maturity Stages of Oocyte

The cavity of the ovary is known as ovocoel. The innermost layer of the ovary is the germinal epithelium, which projects into the ovocoel in the form of lamellae. The ovigerous lamellae that projects into the ovocoel are the seat of maturation of the oocytes.

The germ cells which give rise to the egg are known as the Oogonia, which develops into a ripe egg or ovum, by the process of oogenesis. An oogonium passes through several maturation stages before it develops into a ripe egg. These stages are known as oocytes. The different developmental stages of the maturation of oocyte can be classified as:

I. Primary Oocyte Growth

It is characterised by the growth or increase in size of the immature oocyte and formation of a follicular layer.

1. Oocyte 1

- ☆ It is slightly larger than oogonium
- ☆ Nucleus with 2-3 nucleoli

2. Oocyte II

- ☆ Increase in size of the oocyte
- ☆ Appearance of several small nucleoli along the periphery of the nuclear membrane
- ☆ Presence of a yolk nucleus close to the nuclear membrane.

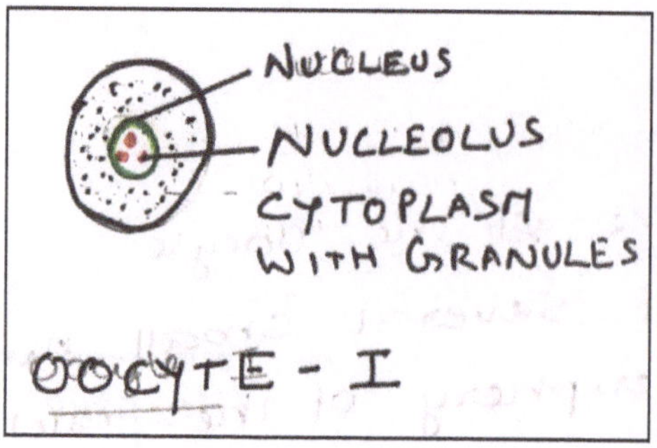

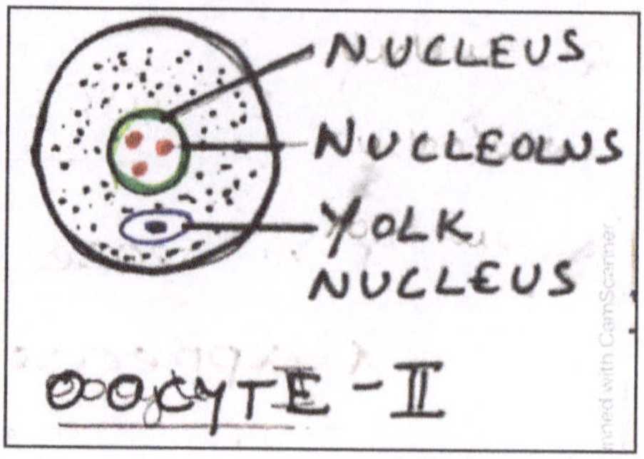

3. **Oocyte III**

- ☆ Further increase in size of the oocyte
- ☆ Appearance of a thin layer of follicular cells around the cytoplasm.
- ☆ Nucleii passes out of the nuclear membrane into the cytoplasm.

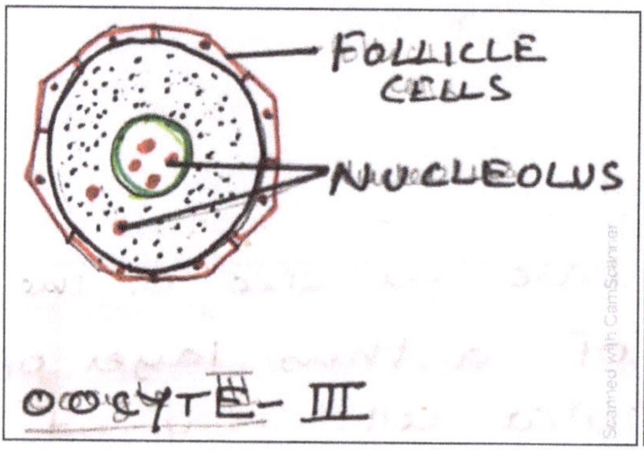

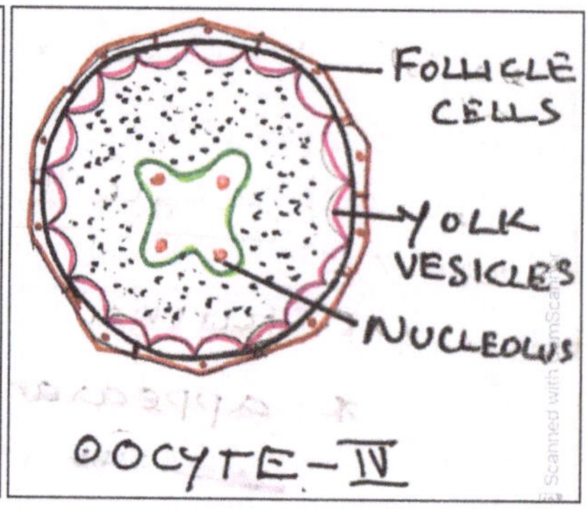

II. Yolk Vesicle Formation

characterised by the formation of vesicles containing a glycoprotein material and formation of Zona radiata

4. **Oocyte IV**

- ☆ Increase in size
- ☆ Presence of a number of small, clear, vacuoles called the yolk vesicles, along the periphery of the ooplasm
- ☆ Nuclear membrane becomes wavy, nucleoli enter into pockets of nuclear membrane and pass out into ooplasm.

5. **Oocyte V**

- ☆ Increase in the number of yolk vesicles which fills the entire ooplasm

- ☆ A vitelline membrane or Zona radiata is seen between the ooplasm and follicular layer (Zona granulosa)
- ☆ Extrusion of nucleoli continues

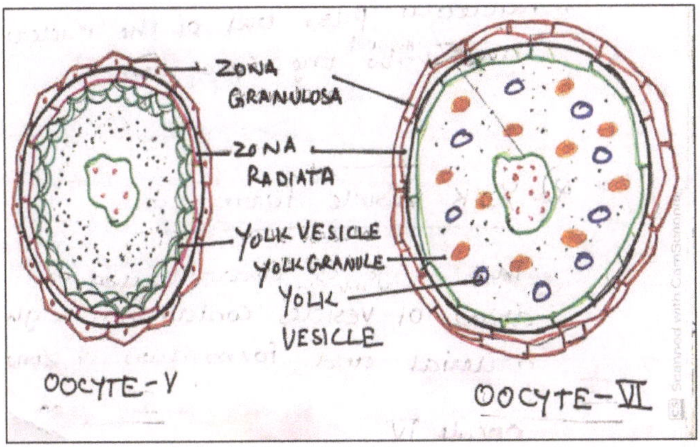

III. Vitellogenesis

It characterised by the synthesis of yolk within the egg cell.

6. Oocyte VI

- ☆ Minute yolk granules appear in the ooplasm and accumulate in the peripheral region
- ☆ Yolk granules gradually fill the entire ooplasm and the smaller ones fuse to form large sized yolk globules
- ☆ Oocyte increases in size and a thin layer of fibroblasts, called theca can be seen external to the follicular layer

IV. Final Maturation

It is a very short stage characterised by the preparation of the egg for ovulation.

7. Oocyte VII

- ☆ The yolk vesicles fuse and becomes large
- ☆ The yolk granules increase in size to form large yolk globules, which further increase in size due to the deposition of more yolk
- ☆ The yolk vesicles are pushed to the periphery to form a layer of cortical alveoli
- ☆ The nucleus moves towards the periphery and finally breaks down.

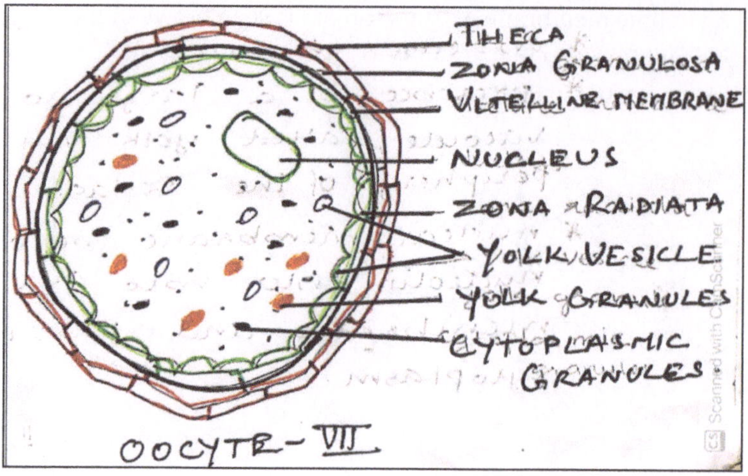

The Ripe Egg

- ✰ Yellowish in colour, translucent
- ✰ Contains large amount of yolk globules
- ✰ Yolk vesicles lie scattered in the cytoplasm
- ✰ Nucleus not visible.
- ✰ Egg is surrounded by an outer layer of theca, followed by a layer of follicular epithelium, the Zona Granulosa and an innermost layer of Zona Radiata
- ✰ Follicular epithelium is responsible for the deposition of yolk in developing ova and it also secretes ovarian hormones.

Spermatogenesis

The spermatozoa is an elongated structure, slightly conical or cylindrical in shape, with radial symmetry. It consists of a head, mid piece and a flagellum.

The **head** consists of an acrosome and nucleus. The head is elongated with an umbrella shaped acrosome and an elongate nuclear region. The acrosome consists of an enzyme complex for penetrating the egg layers and may not be present in all teleosts.

The **middle piece** is small, contain centrioles and mitochondria to provide energy for movement of sperms.

The **flagellum** or **tail** consists of a cytoplasmic membrane and an axoneme, comprising of microtubules.

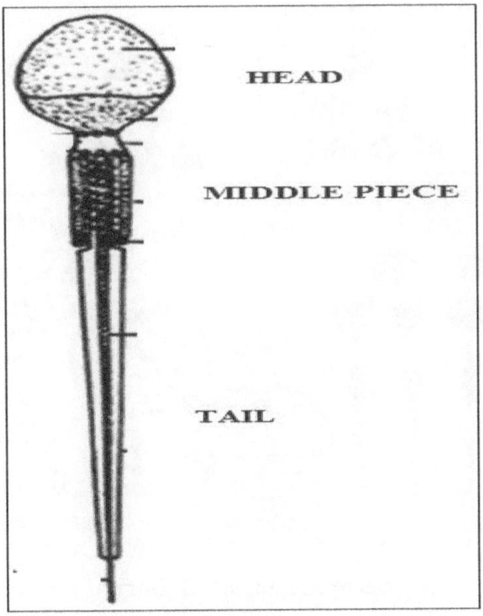

Teleost Sperm.

Spermatogenesis

The formation of spermatozoa from sperm mother cells or spermatogonia through a series of cytological changes is known as **spermatogenesis.** During spermatogenesis, the spermatogonial cells are transformed from undifferentiated diploid (2n) cells into highly specialized haploid (n) spermatozoa.

The three stages of spermatogenesis are;

1. The spermatogonial phase/spermatocytogenesis
2. The meiotic phase/meiosis
3. The spermatid phase/spermiogenesis.

The **spermatogonium**, in the early stages in fishes is a large oval cell with one large and round nucleus. In the spermatogonial phase, spermatogonia proliferate through repeated mitotic divisions and grow, to form a primary spermatocyte.

In the second stage, the **primary spermatocytes** undergo reduction division to form **secondary spermatocytes** and these cells, after the second meiotic division, gives rise to **spermatids.**

The spermatids metamorphose into motile and functional male gametes, the sperms or spermatozoa. The process of metamorphosis of spermatids into sperms is called **spermiogenesis.**

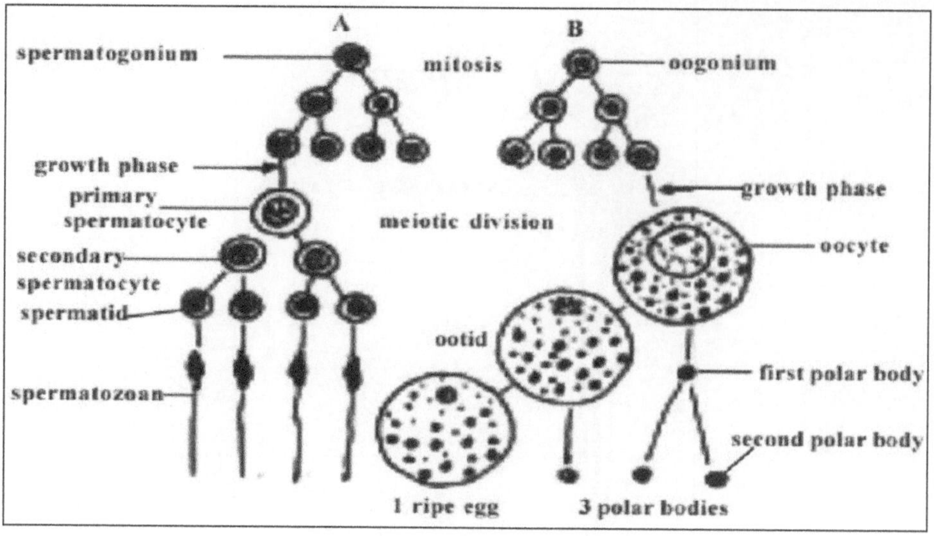

A: Spermatogenesis; B: Oogenesis

Spermatogenesis takes place within the seminiferous tubules of the testes and is synchronous throughout the testis. Prior to spawning, the seminiferous tubules filled with mature sperms rupture to release their contents, either into the spermatic duct or the body cavity.

Cortical Alveoli:

The cortical alveoli or cortical vesicles or yolk vesicles are conspicuous cytoplasmic structures in teleost oocytes. The cortical alveoli content is released into the perivitelline space at the moment of fertilization, blocking polyspermy and contributing to chorion hardening.

Chapter 5

Hormones Involved in Reproduction in Teleosts Fishes

Different hormones are found to play an important role, in teleosts, in the maturation of gonads and release of gametes. The important among these are

1. Gonadotropin Releasing Hormones (GnRH)

The neurohypophysis is made up of neurosecretory cells, whose cell bodies are located in the hypothalamus. These cells can respond to signals from the brain. The cell bodies of neurohypophyseal neurons are arranged in distinct groups or nuclei within the hypothalamus. The two important nuclei or aggregations of cell bodies in the hypothalamus are

a) Nucleus pre-opticus (NPO)

b) Nucleus lateralis Tuberis (NLT)

Hormones produced by NPO neurons (vasopressin and oxytocin) are directly released into a blood channel running between neurohypophysis and adenohypophysis.

The axons, whose cell bodies make up the Nucleus Lateralis Tuberis end directly on gonadotrops of the PPD, in the adenohypophysis. The chemical messengers released at the junction between the axonal terminals of the NLT and gonadotrops are known as **Releasing Hormones.** These hormones stimulate the production of gonadotropins and their subsequent release into the blood stream.

Corresponding to the two gonadotropins, the fish hypothalamus produces two types of **Gonadotropin Releasing Hormones.**

a) Follicle Stimulating Hormone Releasing Hormone (FSH-RH)

b) Luteinising Hormone Releasing Hormone (LH-RH)

Gonadotropin Releasing Hormones in fishes are similar to those of mammals. They are small 10 amino acid peptides. Only mammalian and Salmon Gonadotropin Releasing Hormones are used in routine induced maturation in fishes.

2. Gonadotropins

These are hormones that stimulate the growth and development of gonads. They are glycoproteins (proteins with attached sugar molecules) produced by the pituitary gland. The fish pituitary produces two types of gonadotropins – **Follicle Stimulating Hormone (FSH)** and **Luteinising Hormone (LH).** FSH is rich in carbohydrates and influences vitellogenesis, production of steroid hormones by gonads and spermatogenesis.

LH is low in glycoproteins and is involved in the control of oocyte maturation and ovulation.

3. Gonadotropin Releasing Hormone Inhibitory Factor (GnRIF)

The release of gonadotropins by pituitary in response to Releasing Hormones was found to be inhibited by another neurohormone from the hypothalamus, called **Dopamine.** Dopamine interferes with the action of releasing hormones **on gonadotropin** cells or gonadotrops, thus inhibiting the production of gonadotropins. The action of Dopamine prevents the maturation of gonads.

4. Gonadal Steroids/Sex Steroids

Hormones produced by the gonads play an important role in regulation of reproductive cycles.

a) Testes

The steroid hormones produced by the testes are the **androgens.** They are C_{19} steroids (compounds containing 19 carbon atoms). The important androgens produced by the testes are **testosterone** and **11-keto testosterone**. The androgens are produced by the interstitial cells of the testes.

b) Ovaries

In the ovary, the oocytes are surrounded by a complex, multilayered follicle, made up of a number of cell layers- an inner granulosa layer separated by a basement membrane from the outer thecal layer. Gonadotropic hormones stimulate the production of testosterone by the thecal layers, which are transferred to the granulos layer and converted into oestrogens by the enzyme **Aromatase.**

The primary hormones produced by the ovaries include oestrogen and progesterone. The oestrogen travels to the liver through the blood stream and stimulate the production of vitellogenin or yolk protein. The vitellogenin travels back to the ovary through the blood stream and enters the oocyte, in which it is stored in the form of yolk globules. The uptake of vitellogenin is facilitated by the gonadotropins, FSH and LH.

The other hormone produced by the ovaries – **progesterone** is secreted by the granulosa cells of the ovarian follicles. Progesterone causes an increase in oocyte diameter, due to water uptake in the cytoplasm.

Chapter 6

Principles of Induced Maturation and Breeding

Induced breeding is a process by which organisms are stimulated to breed or spawn in captive conditions by:

1. Natural or synthetic hormones
2. Providing suitable environmental conditions.

This stimulation results in timely release of eggs and sperms from ripe gonads.

Certain environmental parameters like photoperiod, rain, temperature, water current *etc.* influence the hormonal activity from pituitary and gonads. Insufficient release of hormones in captive conditions or in a culture environment may occur, due to:

1. The non availability of the required environmental factors or
2. The environmental factors may not persist for a sufficient length of time, for spontaneous maturation to occur.

Induced spawning by hormones stimulates the surge of gonadotropins thus bypassing the need for environmental factors.

The Hypothalamus- Hypophyseal- Gonadal Axis in Reproduction

The nervous and endocrine system of fish act together to coordinate reproduction. The chain of events, commences from the environment and terminates with the release of gametes.

1. The reception of environmental stimuli such as photoperiod, temperature, amount of rainfall *etc.* is processed by the sensory receptors and information is passed to the brain.
2. The neural information, upon reaching the hypothalamus of the brain influence the pituitary by releasing chemical messengers, known as releasing hormones.
3. The releasing hormones stimulate the pituitary to produce hormones known as Gonadotropins, which are released into the circulation and whose target organs are the gonads.
4. The gonadotropins influence the production of sex steroids in the gonads, which are responsible for gamete maturation and release of gametes under favourable conditions.

This series of events, leading to spawning is known as the **hypothalamus-hypophyseal- gonadal axis.**

Environmental stimuli

↓

Brain

Neural connections

↓

Hypothalamus

Releasing Hormones

↓

Pituitary

Gonadotropins

↓

Gonads

Sex steroids

↓

Maturation and Release of gametes

Hypothalamus – Hypophyseal – Gonadal Axis.

This information provides the fish culturists with four different options or levels of induced maturation in fishes. These are.

1. Manipulation of environmental conditions or creating conditions conducive for breeding

2. Administration of Releasing Hormones
3. Administration of Gonadotropins
4. Administration of Sex Steroids.

1. Environmental Control of Reproduction in Fishes

Reproduction in culture fishes is restricted to a specific period of the year when environmental conditions are favourable. The gonadal maturation of finfishes is basically controlled externally by environmental conditions and internally by the endocrine system. Induced maturation by the manipulation of environmental conditions are the most desirable, since, the subsequent steps of the hormonal chain are taken up by the body itself. The important environmental factors which control the gonadal maturation of finfishes include:

1. Photoperiod
2. Temperature
3. Flooding of water
4. Food availability
5. Stress.

Of these photoperiod and temperature are the main factors influencing reproduction. In nature, gonadal maturation and spawning in most of the finfish species are regulated by manipulation of day length or temperature and providing optimum quantity of food.

Photoperiod

In finfish species, which spawn in spring or early summer, gonadal maturation is stimulated by long photoperiods, particularly in combination with warm temperatures. The photoperiod should be increasing gradually over a period of time.

e.g. Brook Trout (*Salvelinus fontinalis*)

Rainbow trout (*Salmo gairdneri*)

In species which spawns during autumn or early winter, gonadal development is favoured by short or decreasing photoperiods. A decreasing photoperiod is much more effective than a constant short photoperiod in stimulating gametogenesis. *e.g. Onchorhynchus nerka.*

Photoperiod may involve day length, change in day length, spectrum of light or light intensity. The pineal and eyes are the organs which senses changes in the photoperiod.

The photoperiod may also depend on the intensity of light and light intensity plays an important role in gonadal maturation of the milk fish *Chanos chanos.*

Temperature

In majority of finfishes there is a maximum and minimum limit of temperature range within which the fishes breed. The critical temperatures above and below which the fish will not reproduce vary from species to species.

Long photoperiods stimulate gametogenesis in some species like *Gambusia affinis*, only in combination with warm temperatures.

In some other fish species low temperatures promote sexual matutration, regardless of photoperiod, although the effect is enhanced by short photoperiods.

Temperatures influence gonadal stimulation by:

1. Direct action on gametogenesis
2. Action on pituitary gonadotropin secretion
3. Action on responsiveness of liver to oestrogen, during vitellogenesis.
4. Action on responsiveness of gonad to hormone stimulation.

Salinity

The fluctuations in salt content of the water medium have been found to have an effect on gonadal development.

In estuarine species increase in salinity has been found to prevent gonadal maturation.

Other Factors

Influencing gonadal maturation are oxygen content, pH, food availability, stress and pollutants.

An increased oxygen concentration accelerates or induces ovulation in carps.

Reproduction in pup fishes (*Cyprinodon neradensis*) in inhibited, when pH is lowered from the normal habitat level of 8.3.

Decrease in food availability was found to suppress gonadal development in *Salmo gairdneri* and *Salmo trutta*. Improved feeding accelerated maturation time in the flounder, *Pleuronectes* spp and the herring, *Clupea herengus*.

In the temperate golden perch, *Plectoplites ambiguous*, yolk laden oocytes do not mature in the absence of floods.

Stress is another factor affecting gonadal maturation in fishes. Handling stress has been recognized as responsible for the failure in induced maturation of milk fish.

Pollutants such as heavy metals, pesticides and oils are environmental factors inhibiting gonadal activity.

Environmental Control of Reproduction in Penaeid Prawns

As in fishes penaeid prawns can be made to spawn in captivity by the manipulation of environmental conditions. The adult penaeid prawn matures and breeds in seawater conditions; hence seawater parameters are of utmost importance in breeding. The different sea water parameters include

1. **Salinity**: should be maintained between 28 – 36 ppt, matching that of oceanic waters in which natural breeding occurs.
2. **Temperature:** The optimum temperature for breeding is believed to be 27-29°C. Generally temperatures below 26°C will reduce the reproductive performance. Increasing temperature and day length are usually important cues for breeding.

   Sometimes a mild stimulus can be given to a spawning female by transferring to a tank containing water of slightly elevated temperature (1-1.5°C).
3. **pH**: should be kept between 8 – 8.2. To maintain the pH, frequent additions of NaoH or $CaCO_3$ can be used.
4. **Dissolved Oxygen**: should be maintained close to saturation or at least above 5 ppm. Air diffusers are used in maturation tanks and may be kept discontinued during evening to reduce turbulence and make conditions favourable for mating.
5. **Heavy Metals**: Contamination of seawater may be avoided as well as care should be taken to prevent contamination of seawater systems by construction materials like pipes, pumps *etc.*

Other Environmental Conditions

Apart from seawater conditions, a number of environmental factors or conditions are believed to affect reproduction in captivity. They are

1. **Size and Shape of Maturation Vessel** : Circular tanks with a maximum diameter of 3.65m are recommended. This condition allows the male prawn to chase the females during the courtship and mating process.
2. **Water Depth**: may range from 35 – 50 cm. Very shallow tanks are found to inhibit mating.
3. **Stocking Density**: Common stocking densities for adults in maturation tanks are 3-5 individuals/m^2 and is dependent on the size of the brooder and water quality. Best results in *P.monodon* is achieved with stocking densities of less than 3/m^2.
4. **Sex Ratio**: Male to female ratios of 1:2 or 1:3 are found to be adequate for mating.
5. **Substrate**: No substrate is usually required except for strong burrowing species like *Penaeus japonicas.*
6. **Tank Colour and Light Intensity**: Black coloured tanks and dim lights were found to be stimulative to spawning.

7. **Photoperiod**: 14-16 hours light period is actually preferred for spawning.
8. **Water Exchange Rate**: is dependent on conditions like animal size, stocking density, feed type and frequency *etc.* A 100 per cent to 400 per cent water exchange daily is ideal for sea water and water is usually introduced below the surface to reduce turbulence.
9. **Nutrition**: is of utmost importance in maturation. Diets used for reproduction consists of one or more fresh or fresh frozen marine organism ingredients like Squid, Mussel, Clam, shrimp, polychaete worms *etc.* They are chopped and fed 2-5 times daily. Along with these, a high quality commercial formulated diet is often fed at the rate of 25 per cent of the total dry weight of the feed.

2. Use of Gonadotropin Releasing Hormones in Induced Maturation

Releasing hormones are produced by the Nucleus Lateralis Tuberis of the hypothalamus and they act at the hypothalamus pituitary interface. Releasing hormones convert the neural signals into hormonal signals.

The two important Releasing hormones used in induced breeding are:

1. Mammalian Gonadotropin Releasing Hormone
2. Salmon Gonadotropin Releasing Hormone

Both are 10 amino acid peptides, with an amino group attached to Position 10.

Because the releasing hormones are easy to synthesise, several analogues (structurally similar compounds) were prepared by inserting amino acids at different positions. The advantages of these analogues are;

1. They are more potent and superactive than the original Releasing Hormones.
2. They are more stable and resistant to enzymatic breakdown.

Two analogues of releasing hormones used in fish culture are:

1. Mammalian gonadotropin releasing hormone analogue (GnRHa), also called Luteinising hormone releasing hormone analogue (LH-RHa)
2. Analogue of Salmon GnRH called Salmon Gonadotropin Releasing Hormone Analogue (SnGnRHa)

These analogues are prepared by:

1. Substituting with D-amino acids, at the 6th position.
2. The 10th amino acid is replaced with an ethylamide group

Advantages of Releasing Hormones in Fish Culture

1. They act early in the hormonal chain

2. They are simple molecules and easily manufactured
3. They are not species specific
4. More effective and super active analogues can be prepared
5. They are active at low concentrations
6. They are economical

Availability and Storage

GnRHa is available as a dry powder. Some are mixed with an inert binder, for easy handling of small quantities. For storage, gonadotropin releasing hormone solution, it should be frozen as a concentrated stock solution in distilled water.

3. Use of Gonadotropins in Induced Maturation

The different gonadotropins used in induced breeding are

1. Placental gonadotropins
2. Pituitary Gonadotropins

1. Placental Gonadotropins

They are produced by the mammalian placenta. The most common placental gonadotropins used in induced breeding are

a) Human Chorionic Gonadotropin (HCG)

It is secreted by the placenta in mammals. HCG is a glycoprotein composed of 237 amino acids. It promotes the maintenance of corpus luteum during the beginning of pregnancy. This allows the corpus luteum to secrete progesterone during the first trimester of pregnancy. Progesterone provides the uterus with a thick lining of blood vessels and capillaries, to sustain the growing foetus.

HCG is excreted in human urine during pregnancy and it can be extracted from the urine. The secretion of chorionic gonadotropins commence from the early stages of pregnancy and reaches a maximum between 50-60 days. The level of HCG gradually declines upto 80 days. After this the levels of HCG remains steady and again fall towards the close of pregnancy and disappear with the expulsion of the foetus and placenta.

> **Corpus luteum:** or yellow body is a temporary endocrine structure in female ovaries and is involved in the production of high levels of progesterone. It is the remains of the ovarian follicle that has released a mature ovum during a previous ovulation.

A very rapid growth of ova in an under matured female fish and their spawning can be observed when the fish is injected with the urine of pregnant women, which contains gonadotropins.

HCG behaves like the luteinizing hormone, when injected in fish. The effective dose of HCG varies from 100- 2000 International Units/kg weight of the fish. Crude and centrifuged or partially purified human pregnancy urine, when used in combination with fish pituitary extract, produced successful ovulation and spawning of Indian major carps.

Disadvantages

1. Since HCG varies much from fish gonadotropins, some species of fishes require high doses and some may not respond at all.
2. This method is not very practical or economical

Highly purified and potent preparations of HCG available in the market include

1. Prolan (CIBA, ORGANON, BAYER)
2. Antuitrins (PARKE DAVIS)
3. Gonatropin (TECKOKUZOKI Company Ltd)
4. APL (AYERST LABS)
5. CHORIOGONIN (A trade name for chorionic gonadotropin used by Russians, first time in Loach)

b. Pregnant Mare Serum

1. Equine Chorionic Gonadotropin or Pregnant Mare Serum Gonadotropin

It is produced in the chorion of pregnant mares (female horse). It is used in combination with progesterone to induce ovulation in livestock, prior to artificial insemination. It stimulates production of both FSH and LH. It enhances the duration of spawning process in carps.

2. Pituitary Gonadotropins

Pituitary gonadotropins are produced by the pituitary gland.

a. Mammalian Pituitary Hormones

Luteinising Hormone isolated and purified from mammalian pituitaries is found to be a potent agent in causing ovulation and spawning in females and spermiation in males.

Human Chorionic Gonadotropin is often used in combination with mammalian pituitary extract. The preparation is called Synahorin and is used in the induced spawning of grey mullets and Indian major carps.

Puberogen is another hormone preparation used in the breeding of sea bass (*Lates calcarifer*). It consists of 63 per cent Follicle Stimulating Hormone and 34 per cent Luteinising Hormone.

b. *Partially Purified Fish Gonadotropins*

The amount of gonadotropin content in crude pituitaries cannot be estimated correctly. The problem is overcome by purification of the pituitary extract. Partially purified salmon gonadotropin (SG G 100) is a commercially available hormone and has a long storage life.

Using a Gonadotropin Releasing Hormone Analogue with a Dopamine Antagonists (LinPe method)

Gonadotropin secretion in fishes was found to be controlled by 2 factors. It is stimulated by releasing hormones and inhibited by dopamine. Inhibition by dopamine can be overcome by using a drug that blocks the action of dopamine. Such compounds are called **Dopamine Antagonists**.

e.g. Pimozide and Domperidone are the most common. Other examples are Spiperone and haloperidol. Injecting a gonadotropin releasing hormone analogue followed by or in combination with a Dopamine Antagonist is called the LinPe method (after Lin and Peter)

If used separately, Domperidone or pimozide can be used as a suspension of 0.7 per cent NaCl solution containing 0.1 per cent metabisulphate as an antioxidant. They can also be dissolved in propylene glycol and injected. GnRHa is available as a dry powder.

Ovaprim

Ovaprim is a combination of an analogue of Salmon Gonadotropin Releasing Hormone and a dopamine antagonist in a stable solution. It can be used in induced breeding as a single injection. It can be stored in ambient temperatures, even in the tropics, for more than a year. It is packed in 10 ml and 100 ml vials. It is the first ready to use product for inducing ovulation and spermiation in fishes.

Ovaprim was developed and marketed by Syndel Laboratories, Canada. In India, it is distributed by Glaxo. It has been found to be effective for a large number of freshwater and marine fishes.

4. Administration of Sex Steroids in Induced Breeding

Gonadotropins stimulate the biosynthesis of progesterone in the ovarian follicles, which induces final maturation of the ova. Estrogen plays a role in the earlier stages of oocyte development, particularly vitellogenesis.

There are two methods of using sex steroids. They are:

1. Direct Use of Sex Steroids

Induction of oocyte maturation and ovulation in fishes have been achieved by the direct administration of progesterone

e.g. Goldfish, Rainbow trout

2. Using Antioestrogens (or by feedback control of gonadotropin secretions)

Production of sex steroids by the gonads is controlled by a feedback system. A rise in the level of steroids causes a decrease in gonadotropin secretion, which lowers the level of steroid production. The pituitary contains steroid binding sites and secretion of gonadotropins is lowered when steroids are bound to these sites.

By using a compound that prevents the gonadal steroids from binding to the pituitary, a continuous production of gonadotropins and sex steroids can be ensured.

Such compounds, for *e.g.* **Clomiphene citrate** and **Tamoxifen** are known as Antiestrogens and can be used in induced breeding.

Induction of ovulation in goldfish has been achieved by the implantation of Clomiphene in the pituitary and the release of gonadotropins in common carps has been achieved by dosages of clomiphene in the range of 1 mg/kg of body weight.

Hypophysation

Hypophysation refers to the induced breeding of fishes by the injection of pituitary extracts. It contains gonadotropins, FSH and LH, which leads to the maturation of gonads and spawning in fishes. Hypophysation involves the following steps.

1. Selection of Donor Fish

The fish from which pituitary is collected is known as the donor fish. For the selection of a donor fish, the following factors are considered.

a. Stage of Maturity

The pituitary gland should be collected from fully ripe or gravid fishes. The potency of the gland decreases after spawning and those collected from immature fishes have low levels of gonadotropins. Glands from donors collected within 6-8 hours of spawning have also been found to be effective in breeding.

b. Condition of Donor Fish

Freshly killed fishes are the best donors of potent pituitaries. Pituitaries can also be collected from fishes preserved in ice or kept in cold storage at 1- 3°C for 24 hours.

c. Source of the Pituitary Gland

The donor fish can be either male or female. For best results, glands from the same species or closely related species should be used as far as possible.

Common carp can be used as a universal donor, since its pituitaries are found to be effective for other species. Catfish pituitaries can also be used in the induced breeding of carps. Pituitaries of the same species are known as **Homoplastic pituitaries** and those of different species are known as **Heteroplastic pituitaries**.

2. Methods of Collection of Pituitary Gland

The pituitary gland is collected by:

a. *Dissection and Removing a Part of the Brain Case*

A portion of the brain case is cut from the top obliquely, using a sharp knife to expose the brain. The olfactory and optic nerves at the anterior region of the brain are detached by tweezers and the brain is lifted up and laid back. The pituitary lies in a depression, the **Sella Turcica**, separated by a thin membrane from the brain. The membrane is lifted up and the pituitary is removed by tweezers.

b. *Collection of Pituitary Gland through the Foramen Magnum*

If cut pituitaries are available, the foramen magnum is exposed by removing the first vertebrae from the occipital condyles. The olfactory and optic nerves are detached and the brain is lifted up and removed through the foramen magnum. The exposed pituitary is lifted up using tweezers.

3. Preservation of Pituitary Gland

The collected pituitary glands are preserved using absolute alcohol, acetone or by immediate freezing.

In alcohol preservation, the pituitary glands are stored in absolute alcohol in glass vials for 24 hours. After 24 hours, the alcohol is changed and the vials are stored in a cool dry place in a desiccator containing anhydrous calcium chloride.

In acetone preservation, the glands are put in fresh or dry ice (solid Carbon dioxide) chilled acetone, in vials and stored in a refrigerator for 36 hours. The acetone is changed every 12 hours. After the glands are taken out of acetone, they are weighed and stored in fresh vials in a refrigerator.

The alcohol and acetone function as dehydrating and defattening agents respectively.

The pituitary gland may also be frozen immediately after collection and kept in frozen storage in a refrigerator or insulated cans containing dry ice.

4. Preparation and Preservation of Pituitary Extract

The pituitary extract is prepared by homogenising the pituitary gland and subsequent centrifugation or by using Trichloroacetic acid.

The glands are taken in a tissue homogeniser with little quantity of distilled water or 0.3 per cent NaCl solution and macerated. The solution is centrifuged and the supernatant solution containing the pituitary extract can be preserved in

a. Glycerine
b. Propylene glycol

The pituitary extract prepared in distilled water is preserved in pure glycerine in air tight bottles for 3-4 days to allow the sediments to settle down. 20-40 mg of the gland is preserved in 1 ml of distilled water and glycerine taken in the ratio 1: 2. After the sediments settle down, the solution is filtered through a filter paper and is stored in glass ampoules under refrigeration.

Pituitary extracts can also be preserved in propylene glycol and stored under refrigeration.

Pituitary Extraction in Trichloroacetic Acid (TCA)

The Gonadotropins in the pituitary may be extracted by keeping the gland immersed in 1.5 per cent TCA solution for 6-12 hours or using 2.5 per cent TCA solution for 3 hours. The extract is stored under refrigeration.

5. Selection of Breeders

Selection of healthy breeders is a prerequisite for successful breeding. Males and females are stocked separately for a few months before breeding and fed rice bran and oil cake in the ratio 1:1 @2 per cent of the body weight.

Indian major carps weighing 1-1.5 kg are preferred for breeding. Males can be distinguished by the roughness on the dorsal surface of the pectoral fins in the breeding season and females by their soft and bulging abdomens and their swollen, reddish vents. Maturity can be assessed by catheterization (inserting a glass tube into the ovary and drawing up a sample of the ova)

6. Injection of Pituitary Extract

The females are given two injections. The first one is known as **Provocative** or **Preparative** or **Stimulating** dose and consists of 2-3 mg of pituitary/kg body weight of the fish. After an interval of 6 hours, the resolving dose of 2-3 mg of pituitary/kg body weight of the fish.

The males are given only a single injection of dosage, 2-3 mg/kg body weight of the fish at the time of the second dosage of the female.

The injection is usually intramuscular or and given into the muscles, below the dorsal fin or below the lateral line of the caudal peduncle. After injections, the male and female breeders are introduced in the ratio, 2: into the breeding hapa.

Chapter 7

Anaesthetics

Anaesthetics are chemicals used to reduce stress on the fish and minimise its body activities during handling and transportation. Handling of the fish is required during routine fish husbandry practices as well as for induced spawning, surgical and diagnostic purposes.

Anesthesia is generally defined as a condition caused by an applied external agent, resulting in a loss of sensation through depression of the nervous system. Anesthetics may be local or general, depending on their application.

The different stages of anaesthesia induction in an animal involve;

1. **Sedation:** Motion and breathing is reduced
2. **Anaesthesia:** Partial loss of equilibrium, but reacting to touch stimuli
3. **Surgical Anaesthesia:** Total loss of equilibrium and no response to touch stimuli.
4. **Death:** Stoppage of breathing and heartbeat.

In fishes, the different stages of anaesthesia and recovery are as below.

Stages of Anaesthesia

1. **Stage I:** Loss of equilibrium
2. **Stage II:** Loss of gross body movements but with continued opercular movements
3. **Stage III:** As in Stage II with cessation of opercular movements

Stages of Recovery

1. **Stage I:** Body immobilized, but, opercular movements just starting
2. **Stage II:** Regular opercular movements and gross body movements beginning.
3. **Stage III:** Equilibrium regained and pre anaesthetic appearance.

The condition of the fish or state of anaesthesia achieved usually depends on the length of the fish, the dose and time of exposure. Fishes to be injected for induced breeding need to be sedated enough for weighing, biopsy and injection.

Administration of Anaesthetics in Fishes

Administration of anaesthetics can be done by:

1. Bath Immersion

It involves placing the fish in an anaesthetic bath. After the required procedure, the fish is placed in fresh and running seawater for recovery. The water quality parameters are carefully monitored during the procedure. It is the most widely used method of anaesthesia.

2. Injection

Injection of anaesthetics can be done in larger fishes.

3. Oral Administration

Oral administration is achieved through capsules of food fed to the fish. It is found to be slow acting, probably due to slow absorption by the intestinal tract.

The methods of anaesthesia can be broadly divided into two:

1. **Pharmaceutical methods**: in which anaesthesia is achieved by administration of certain chemicals
2. **Non Pharmaceutical Methods:** in which different methods other than application of chemicals are adopted, like, electric current, hypothermia *etc.*

Pharmaceutical Methods

The different chemicals used for inducing anaesthesia in fishes include.

1. Tricaine Methane Sulphonate (MS 222/Tricaine/TMS)

- ☆ It is the most commonly used anaesthetic in fish culture.
- ☆ It is highly soluble in water, but also more acidic and the solution has to be buffered with NaOH or $NaHCO_3$.
- ☆ It is photosensitive and degrades on standing.
- ☆ It is used @50-100 ppm (mg/L).
- ☆ It is expensive

2. 2-Phenoxy Ethanol

- It is cheaper than MS 222.
- It is a colourless, oily, aromatic liquid, with a burning taste.
- It dissolves in water with shaking.
- It is also used as an antibacterial and antifungal bath.
- It produces anaesthesia at 200 – 500 ppm.
- It is a mild toxin and can cause skin irritation

3. Metomidate Hydrochloride

- It is highly soluble in water
- It acts rapidly at low concentrations (5-20 ppm)
- Recovery time is longer than in MS 222 or 2 phenoxy ethanol
- It is very effective when squirted directly into the mouth of the fish
- It is one of the few anaesthetics, that is effective when injected intramuscularly

4. Benzocaine

- It is related to Tricaine and less stressful to the fish, since it is neutral in solution
- It is available as a white crystalline water insoluble powder that is to be dissolved in ethanol to form a stable solution
- It is available as a water soluble salt, Benzocaine Hydrochloride, but when dissolved in water, it lowers the pH causing irritation to the fish and has to be neutralised.
- Strong aeration of water is necessary due to the hypoxic effect of the drug resulting from depression of the medullary respiratory centres and reduced gill ventilation.
- Since the benzocaine is lyophilic, recovery times and residue levels in body tissues vary according to the amount of stored fat.

5. Quinaldine

- It is not water soluble and hence has to be dissolved in an organic solvent prior to mixing with water.
- Its derivative, quinine sulphate is a water soluble yellow powder, which requires buffering with sodium bicarbonate upon addition to water.
- Its potency is affected by water pH, temperature and hardness and it is less toxic at lower pH and temperature values.
- It provides rapid induction of anaesthesia and recovery.

6. Clove Oil and Derivatives

- ☆ Clove oil is a pale yellow liquid, derived from the leaves, buds and stem of the clove tree (*Eugenia sp.*) Its active ingredients are Eugenol and iso eugenol, comprising 90-95 per cent of clove oil by weight. Eugenol is found to be similar to 2-phenoxyethanol.
- ☆ It is found to be effective in freshwater and marine species at concentrations of 40-120 ppm and at concentrations of 40-60 mg/L for salmonids.
- ☆ It should be dissolved in ethanol, (ratio of 1:9), before mixing into the water.
- ☆ Clove oil has a slightly faster induction time and a longer recovery time than similar concentrations of TMS

AQUI-S

- ☆ Aqui-S was developed by the seafood research laboratory in New Zealand.
- ☆ It is composed of approximately 50 per cent active ingredient, isoeugenol and 50 per cent polysorbate
- ☆ AQUI-S can be dissolved directly into fresh or salt water.
- ☆ It has been shown to be effective at concentrations of 20 mg/L for anaesthetizing juvenile Chinook salmon.
- ☆ The induction of anaesthesia is stress free, since it suppresses the production of cortisol
- ☆ The withdrawal time of the fish from the anaesthetic is nil.
- ☆ It is mainly meant for the transportation of live food fish to the market.

Both clove oil and its derivative AQUI-S, compounds have a wide margin of safety between effective and lethal doses, and fish do not show signs of distress when being anaesthetized.

In addition to all the above certain other chemicals are used much less frequently. They include

1. Metomidate and Etomidate

- ☆ Metomidate is a water soluble powder, which has the properties of a hypnotic, sleep inducing drug
- ☆ Etomidate is a colourless, odourless crystalline analogue of metomidate and propoxate; very expensive and difficult to obtain Metomidate is effective in both fresh and saltwater. The efficient dosages range from 1 - 10 mg/L.
- ☆ A side effect of anesthesia with metomidate is muscle twitching which can make blood sampling difficult.

- ☆ The efficacy of etomidate is pH dependent and it has proven to be more effective in alkaline waters. Temperature influences the toxicity of etomidate, with higher temperatures rendering the drug less toxic.
- ☆ Both drugs are fast acting with induction times of less than 3 minutes and lengthy recovery periods of upto 40 minutes.

2. Propoxate

- ☆ It is a crystalline powder which resembles metomidate and etomidate structurally
- ☆ Freely soluble in both fresh water and salt water.
- ☆ It is stable in solution for long periods.
- ☆ It is 100 times more soluble and 10 times more potent than TMS
- ☆ Higher doses cause respiratory arrest.
- ☆ Effective at concentrations of 0.5 – 10 mg/L

3. Propanidid

- ☆ Pale yellow liquid which is insoluble in water, but soluble in alcohol.
- ☆ Induction and recovery times for propanidid are in the range of 2 - 4 minutes and 5 -10 minutes, respectively, at 1.5 - 3.0 mL/L or for intraperitoneal injections of 2mg/kg in salmonids.
- ☆ Causes no change in blood chemistry (red cell numbers, haemoglobin content *etc.*) of the exposed fish either during or after anesthesia
- ☆ Respiratory and metabolic acidosis, lasting for approximately 1 hour after recovery from anesthesia has been observed.
- ☆ The anesthetic also caused no change in water CO_2 or pH.

* **Respiratory acidosis** is a condition that occurs when the lungs can't remove enough of the carbon dioxide (CO_2) produced by the body. Excess CO_2 causes the pH of blood and other bodily fluids to decrease, making them too acidic. Normally, the body is able to balance the ions that control acidity.

4. Ketamine Hydrochloride

- ☆ White crystalline powder
- ☆ Water solubility of 200 g/L
- ☆ Widely used as an anesthetic both in human and veterinary medicine, and is safe for the handler.
- ☆ Wide safety margin between lethal and effective doses
- ☆ Injectible drug, generally dissolved in saline and administered intravascularly (i.v.) at a dose of 30 mg/kg for salmonids

- Because the drug is an injectible, it is not appropriate for large groups of fish. However, intramuscular injections with a dart gun to specific individual fish in a tank or stream may be a successful application.
- Does not block ventilatory rhythm; hence is appropriate or desirable for long-term anaesthesia, where it is not possible to maintain constant irrigation of the gills with water.

5. Lidocaine (Xylocaine)

- Insoluble in water but freely soluble in acetone or alcohol.
- Generally used in the hydrochloride salt form this is freely soluble in water
- Cardiac depressant used by veterinarians topically (application to a body surface including skin or inside of the mouth)or injected as a nerve block.
- Used in combination with bicarbonate (1g/L) to anaesthetise carp (*Cyprinus carpio*), Tilapia (*Oreochromis mossambicus*) catfish (*Ictalurus punctatus*). Addition of bicarbonate has been demonstrated to enhance the anaesthetic effects of lidocaine.

Non Pharmaceutical Methods of Anaesthesia

Non pharmaceutical methods do not involve the usage of drugs and is used in cases where:

1. Drug withdrawal times may interfere with marketability
2. In field studies and locations, where no approved drugs are commercially available.

Some non-pharmaceutical methods of anaesthesia are:

1. Carbon Dioxide Bubbled through Water

- Carbon dioxide is used as an anaesthetic especially during transport.
- It produces sedation and anaesthesia and leaves no chemical residues in fish
- Carbon dioxide is extremely soluble in water and carbon dioxide solution can be prepared by diffusing the gas into water using a gas cylinder
- A solution of carbon dioxide can also be produced by using sodium bicarbonate and acetic acid. When dissolved in water sodium bicarbonate releases carbon dioxide, if the pH is acidic.

Disadvantages

1. Prolonged induction and recovery periods
2. Lowering of pH: – CO_2 dissolved in water produces carbonic acid, which may lower the pH of the medium and cause irritation to the fish.

3. The apparatus required is impractical and inconvenient for broodstock sedation.

2. Electroanaesthesia

- Anaesthesia can be directly induced using direct or alternating currents
- It is not much effective in sea water because seawater is more conductive than the fish
- Fishes subjected to low voltage DC current become immobile only as long as the fishes are in the electric field, and do not produce true anaesthesia.
- An AC current of strength 220-240 V can produce loss of reactivity to touch stimuli for upto 5 minutes.
- Larger fishes are affected more rapidly than smaller fishes, and the length of time, the fish remains in the anaesthetized state increases with body length.

Disadvantages

- Electric stimulations can produce violent muscular responses that can disfigure or kill
- Electricity is unsafe to use around water and the safety of the animal and operator are at risk.

III. Hypothermia

- Lowering of water temperature tranquilises or immobilizes the fish.
- Lower water temperature reduces the metabolic activity and oxygen consumption of the fish
- Water can be cooled by refrigeration or by adding ice.
- The rate of cooling should be controlled since rapid chilling can produce thermal shock or even death.
- Generally the water temperature should not be reduced more than 1°C for every 15 minutes.
- Hypothermia is often used in combination with chemical anaesthesia, to reduce the amount of anaesthetic required, reduce oxygen consumption and increase the amount of time the fish is anaesthetized.
- Hypothermia has been primarily used for transportation of fishes and crustaceans.

Factors Affecting Anaesthesia in Fishes

Many factors affect the efficiency of anaesthesia in fishes; these can be divided into

1. Biological Factors

- ☆ The rate at which an anaesthetic becomes effective is related to the ratio of gill area to body weight. Larger individuals generally require greater concentration of anaesthetics than smaller fishes, due to relatively lesser gill area.
- ☆ The metabolic rate of aquatic species also affects the rate at which chemicals are absorbed and anaesthesia is induced. For eg: warm water species absorb anaesthetics even in low concentrations and at a faster rate than coldwater species.
- ☆ Many drugs such as MS 222 and Benzocaine, being fat soluble, are retained for a longer time in larger or gravid fishes in their lipid reserves. Hence anaesthesia may last longer and recovery may be slower, in these fishes.
- ☆ Diseased or weakened fishes are much more responsive than healthy fishes to anaesthetic treatment.

Chapter 8

Cryopreservation

It is a branch of cryobiology, which refers to the long term preservation and storage of biological materials at very low temperatures, normally -196°C, the temperature of liquid nitrogen. At this temperature, the cells can be stored in a viable and genetically stable form for a very long time.

Application of Cryopreservation in Fisheries and Aquaculture

1. Cryopreservation makes possible, the almost indefinite storage of desirable genes.
2. The efficiency of selective breeding and stock improvement among cultured animals can be increased.
3. It overcomes the problems of males maturing before females; ie Asynchronous breeding.
4. For wild fish, gamete cryopreservation is the last possibility to preserve unique characters.
5. Conservation of desirable genomes.

The basic principles of cryopreservation include:

1. Freezing
2. Storage
3. Thawing

1. Freezing

During cooling, the cells may be subject to various stresses arising from:

1. Reduction of temperature
2. Physical and mechanical effects of extracellular and intracellular ice.
3. Concentration of extra and intracellular solutions during freezing

The following injuries may occur during the reduction of temperature, in freezing.

During slow freezing, the concentration of extracellular solutes increases and expose the cells to osmotic stress. The extracellular solution has a tendency to withdraw water from the cells and the cell membrane may collapse.

If the cooling rate is fast, ice crystals are formed in the intracellular solution, which injures the cell membrane.

The optimum rate of cooling is therefore one which allows the cooling rate to be high enough to minimize the time of exposure to concentrated solutions and is low enough to minimize the amount of extracellular ice.

2. Storage

After the cells are frozen, they have to be stored safely at ultra low temperatures. Storage temperatures should be -130°C or lower. The commonly used storage temperature is -196°C or the temperature of liquid nitrogen. At these low temperatures, all biological molecules become motionless and therefore cannot participate in any biochemical reactions.

3. Thawing

During thawing, the same physico-chemical processes takes place in reverse and order and the thawing rate should be same as the corresponding cooling rate.

Cryoprotectants

Cryoprotectants are substances which can minimize the damage to cells during the cooling process due to ice crystal formation and solute concentration.

Certain chemicals can increase the balance between the effects of intracellular ice and formation of concentrated extracellular solution, thus improving the survival of the cells during freezing. These substances are termed **cryoprotectants** and they act by

1. Lowering the freezing point of intracellular fluids
2. Binding with elecytrolytes and preventing them from forming lethal concentration.
3. Optimum dehydration.

The main requirements for a cryoprotectant are:

1. Water solubility
2. Non-Toxicity

Cryoprotectants can be classified into two groups based on their ability to penetrate the cells.

I. Permeating Cryoprotectants

Permeating cryoprotectants act by entering the cells and mixing with the intracellular solution. The important and commonly used cryoprotectants are

1. Dimethyl sulphoxide
2. Glycerol
3. Methanol
4. Ethylene Glycol
5. 1, 2 propanediol.

1. Dimethyl Sulphoxide (DMSO)

It is fairly permeable to cell membranes and its toxicity is intermediate between glycerol and methanol. DMSO is a better cryoprotectant than glycerol, because, it enters and leaves the cells faster than glycerol. DMSO concentrations of 5-20 per cent have been successfully used for cryopreservation of sperms of salmon and sea bream.

2. Methanol

It permeates into the cells much faster than DMSO or glycerol. Methanol is highly permeable to cell membranes and also the most toxic. Its toxicity increases at high concentrations, if cooling rate is low.

3. Ehylene Glycol

It is used in the cryopreservation of salmon and trout gametes.

4. 1-2 propanediol or Propylene Glycol

It is also used as a cryoprotectant in certain fishes and is much more effective than DMSO in certain cases.

II. Non-Permeating Cryoprotectants

These cryoprotectants cannot penetrate the cell membrane and act by combining with the extra cellular solution. They include:

1. **Sugars** (Sucrose and Glucose)
2. **Polymers** (Polyvinyl pyrrolidone (PVD))
3. **Proteins** (Egg yolk, serum, skimmed milk and antifreeze proteins found in polar fishes)

Non – permeating cryoprotectants are usually used in combination with any of the permeating cryoprotectants. Some non permeating cryoprotectants used are:

1. **Egg yolk:** It is the most commonly used non – permeating cryoprotectant. It is used at the rate of 5-10 per cent solution.
2. **Skimmed Milk/Milk Powder:** It is used as a cryoprotectant for the sperm of sea bass and can be used at a concentration of 5 per cent.

Extenders

Undiluted gametes are not suitable for freezing and they must be diluted with a suitable extender solution. The function of the extender solution is to inhibit the activation of ova and motility of the sperms. Once the sperm becomes motile, it cannot be reactivated after thawing. Hence care must be taken to prevent the motility of the sperms and activation of ova, before cryopreservation.

An extender solution is a solution of balanced salts and sometimes organic compounds or it is a solution consisting of inorganic and organic chemicals, resembling that of blood or seminal plasma, in which viability of spermatozoa can be maintained during storage.

Extenders include simple solutions containing Sodium Chloride, Sodium Bicarbonate and Lecithin.

Diluted seawater may be used as an extender solution, as in the case of *Mugil cephalus* and some other clupeids.

One of the best known and successful extenders is **Mounib's medium,** with a pH of 7-7.5. It is a solution containing, sucrose, glutathione and potassium bicarbonate.

Dilution Ratio

It is the ratio by volume with which the milt is to be mixed with a Cryodiluent before cryopreservation. **Cryodiluent** is a mixture of cryoprotectant and extender solution.

Usually, the dilution ratio ranges from 1:1 to 1:19 and the optimum dilution ratio is 1:3 (Milt: Cryodiluent)

Equilibration Period

After adding the cryoprotectant, a time interval must be given to the cryoprotectants to penetrate the cell and begin its action, before the medium is taken for freezing. During equilibration period, the cryoprotectant initiates its action by entering the cell.

The equilibration period varies from 30 minutes to 2 hours, and in some cases, no time at all is required.

A generalized cryopreservation procedure includes the following steps:

1. Milt Collection

Milt is collected by stripping and care is taken to wipe the vent dry before milt collection. Milt is to be collected without contamination by urine and faeces.

2. Pre-Freezing

Between milt collection and freezing is a short time period in which, the milt must be kept in frozen storage. This is known as pre freezing. Pre Freezing should be for a short time period only and at reduced temperatures of less than 0°C. It is always desirable to freeze the milt as soon as it is collected.

3. Addition of Cryodiluent

The cryodiluent is usually added in the ratio 3:1. For a 3:1 cryodiluent: Milt ratio, the quantities are

0.15 ml	+	0.39 ml	+	0.06ml	= 0.6 ml
Milt	**+**	**Extender**	**+**	**Cryoprotectant**	**= Milt + Cryodiluent**

The final concentration of DMSO will be 10 per cent

4. Equilibration Period

The equilibration period should be kept to a minimum of less than one hour to prevent cryoprotectant toxicity.

5. Freezing

The already prepared gametes are cooled from 0 to -196°C, during this phase. The commonly used freezing methods are vial, pellet and straw methods.

Polypropylene vials of 1 to 2 ml capacity are used for freezing.

In **straw method**, the diluted semen is drawn into coloured plastic straws and either heat sealed or plugged with a powder, which gels in the presence of liquid to form a gel.

The **French Straws** are thin plastic tubes of 135 mm length, 2- 2.8 mm diameter, having a capacity of 0.25 – 0.5 ml and made up of Polyvinyl Chloride (PVC). It has two ends; factory seal end already provided sealed by the manufacturer, and the laboratory seal end. The laboratory seal end will be open when purchased. It has to be sealed after filling the straw with diluted semen, before freezing. The factory seal consists of two cotton plugs with Polyvinyl alcohol (PVA) powder in between. The laboratory seal is formed with PVA powder or sealing machine, after filling with semen.

In the **Pellet method,** the seminal fluid, mixed with extender solution and cryoprotectant is dropped into small holes which are drilled in dry ice (Solid carbon dioxide, -40°C). The frozen pellets thus obtained can be easily removed for transfer into liquid nitrogen.

6. Frozen Strorage

The vials, pellets and straws are stored in liquid nitrogen at - 196°C.

7. Thawing

The thawing rate is usually as fast and same as the cooling rate, which also helps in post thaw survival and motility induction.

Pellets can be thawed using a thawing solution. Vials can be thawed in a water bath. French straws can be thawed at room temperature.

8. Post Thaw and Insemination

At this stage, the gametes that have survived the cryopreservation procedure are ready for artificial insemination.

Steps in Acryopreservation Procedure

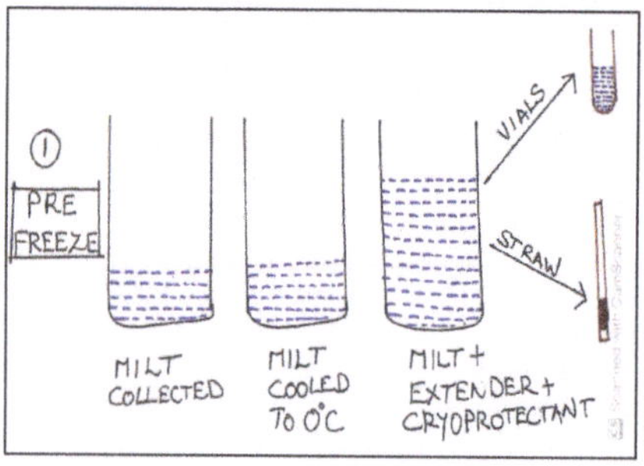

Pre Feeze

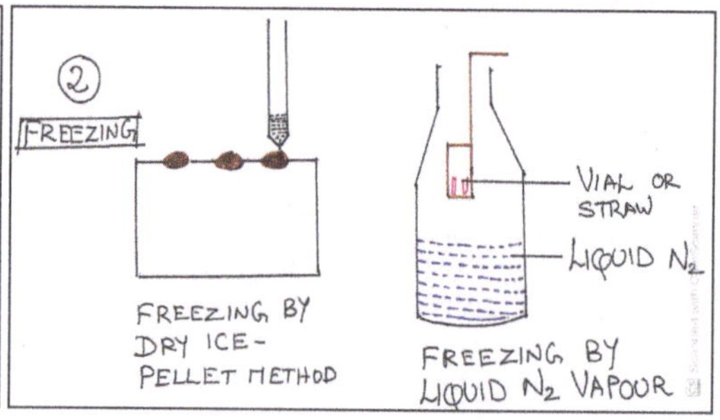

Freezing

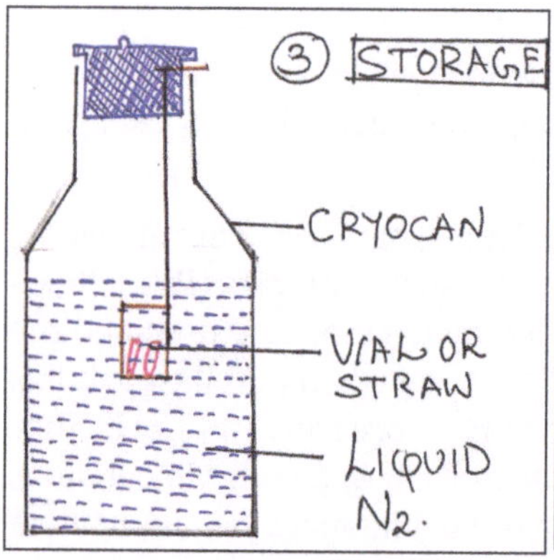

Storage

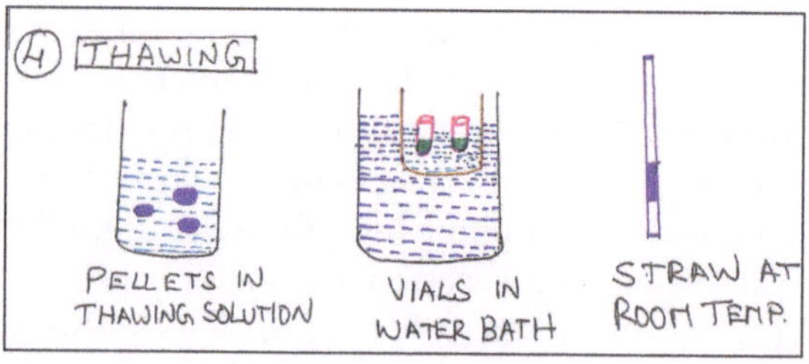

Thawing

Chapter 9

Embryonic Development in Fishes

The developmental stages in fishes can be classified into:

1. Early embryonic development – Starts from fertilisation of the egg and ends with the development of generalised organ systems, when it is transformed into a larva.
2. Transitional Embryonic Development/Larval Stages

   It involves the larval stages in which the organ systems and body forms develop resembling the adult.
3. Post Embryonic development: includes the juvenile and adult phases.

The embryonic development include the following processes:

Structur of a Typical Ovum

The egg is surrounded by a tough layer called **Chorion.** Next to the chorion is the **Plasma** or Vitelline **membrane or** Pellicle. This layer surrounds the yolk and cytoplasm. The cytoplasm of the egg cell is called ooplasm. It contains yolk, pigments and cortical granules. The peripheral layer of the egg cytoplasm is viscous and known as the egg cortex, provided with microvilli formed as outpushings of the plasmalemma. Cortical granules are small membrane bound bodies containing mucopolysaccharides. Embryonic development includes the following processes.

1. Fertilisation and Impregnation

The fertilization in teleosts is external taking place in water, outside the body. During fertilization the pronuclei of sperm and ovum unite together, with the fusion of cytoplasm.

Impregnation is a process by which the egg is penetrated by the sperm. The egg is activated by the entry of the sperm. The sperm enters the egg through the micropyle. In some fishes, the micropyle is funnel shaped. As soon as the sperm penetrates, there occurs a cortical reaction which prevents further entry of sperms.

The micropyle (small pore on the egg surface) is closed by the secretion of large vesicles in the cytoplasm, the cortical alveoli in the egg cortex. The vitelline membrane and chorion is made impermeable to sperm entry. The egg swells up and increases in size. The vitelline membrane separates from the cytoplasm and a perivitelline space appears filled by a perivitelline fluid. After fertilization, the vitelline membrane is known as the fertilization membrane.

After fertilization, the cytoplasm moves towards the animal pole of the egg and forms a cytoplasmic cap. A polarity is thus set up, with the pole containing cytoplasm, known as the animal pole and the opposite pole containing yolk, designated as the vegetal pole. Active cytoplasm is withdrawn both from the surface and from the interior of the egg. As a result the yolk platelets are placed close together to such an extent that the cytoplasm is completely squeezed out from between them. The fertilized egg consists of the cytoplasmic cap on the animal pole, containing the nucleus, the central mass of yolk and a thin layer of cytoplasm surrounding the yolk.

Types of Eggs

Based on yolk content, eggs are classified into

1. **Microlecithal eggs**- Very little amount of yolk is present, distributed evenly in the cytoplasm. (isolecithal)
2. **Mesolecithal egg**: Only moderate amount of yolk is present and concentrated towards the vegetal pole.

   e.g. Cyclostomes and ganoid fishes.
3. **Macrolecithal eggs:** Contains large quantities of yolk, which occupies the entire vegetal hemisphere and major part of the animal hemisphere. The ooplasm and nucleus are displaced towards the animal pole. *e.g.* Teleosts.

Based on distribution pattern of yolk, eggs are classified into:

1. **Isolecithal/Homolecithal eggs:** Very little amount of yolk is present, distributed throughout the ooplasm.
2. **Telolecithal egg:** In mesolecithal and macrolecithal eggs, the yolk is concentrated more towards the vegetal pole. Such eggs are called telolecithal eggs.
3. **Centrolecithal eggs**: Is large and elongate, nucleus is centrally placed, surrounded by a thin layer of cytoplasm. Another thin cytoplasmic layer covers the entire yolk and is connected to the centre by cytoplasmic strands. Eg: Arthropods and Coelenterates.

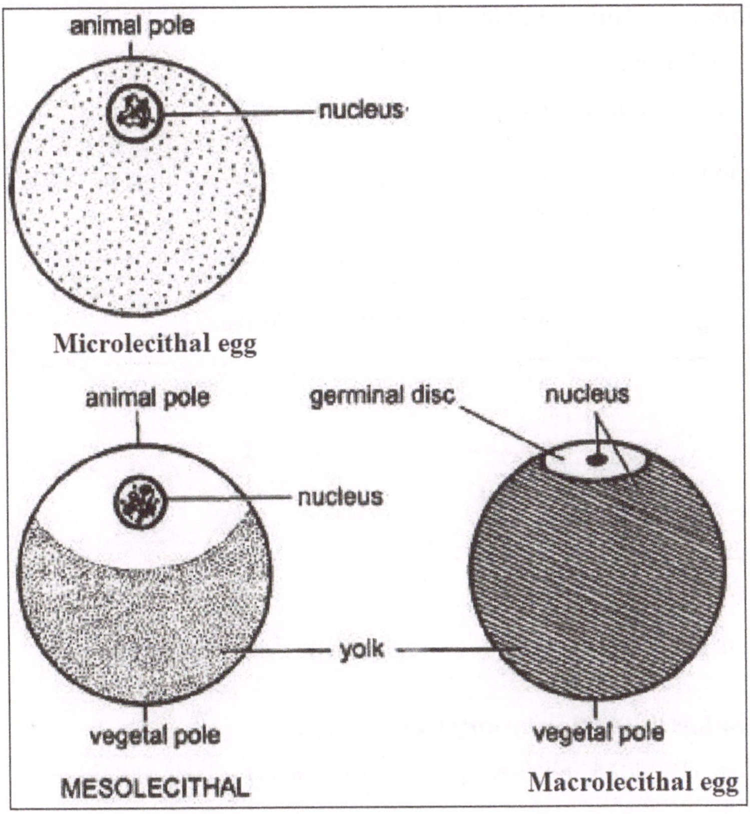

Types of Eggs Based on Yolk Content

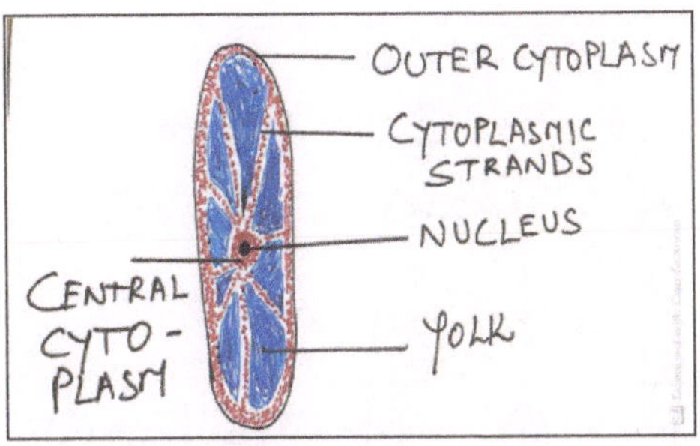

Centrolecithal Egg.

Type of Cleavage

The type of cleavage is influenced by the amount and position of yolk present. It can be classified into:

1. *Holoblastic/Complete Cleavage*

Where the egg is completely divided.

In holoblastic equal cleavage, equal sized blastomeres are formed. In holoblastic unequal cleavage, the blastomeres are unequal in size. The bigger blastomeres are called macromeres and occur in the vegetal pole and the smaller blastomeres or micromeres occur at the animal pole

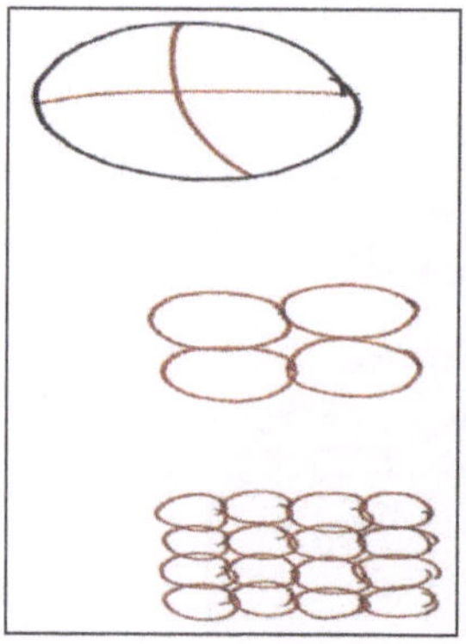

Holoblastic Equal Cleavage.

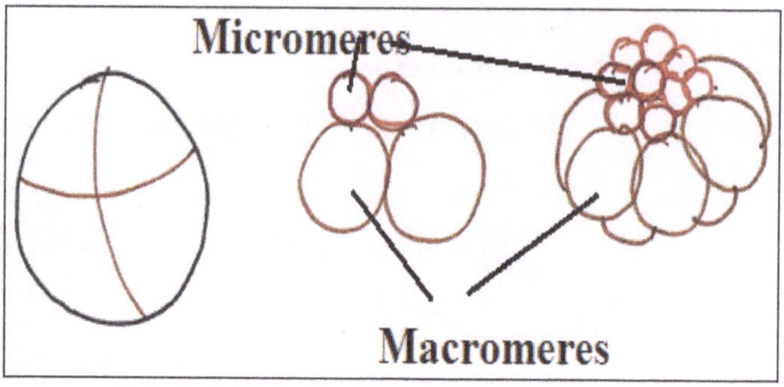

Holoblastic Unequal Cleavage.

2. *Meroblastic/Partial/Incomplete cleavage*

The egg is divided only in the portion of active cytoplasm and the yolky portion remains uncleaved.

In **Meroblastic superficial** cleavage, the cleavage is confined to the peripheral cytoplasm *e.g.* Centrolecithal eggs.

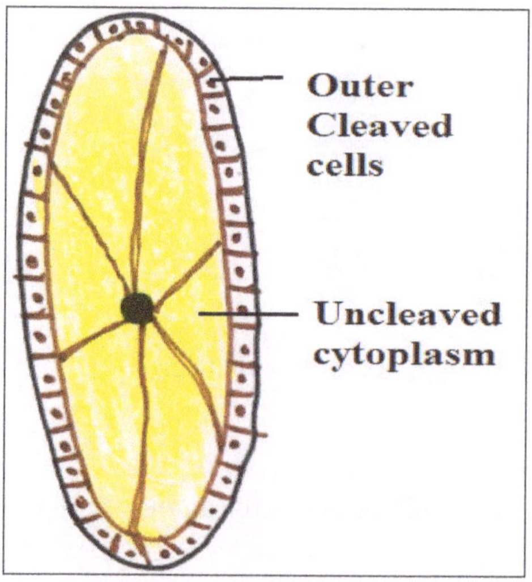

Meroblastic Superficial Cleavage.

In **Meroblastic discoidal** cleavage, the cleavage is confined to a disc shaped area of active cytoplasm at the animal pole, known as blastodisc.

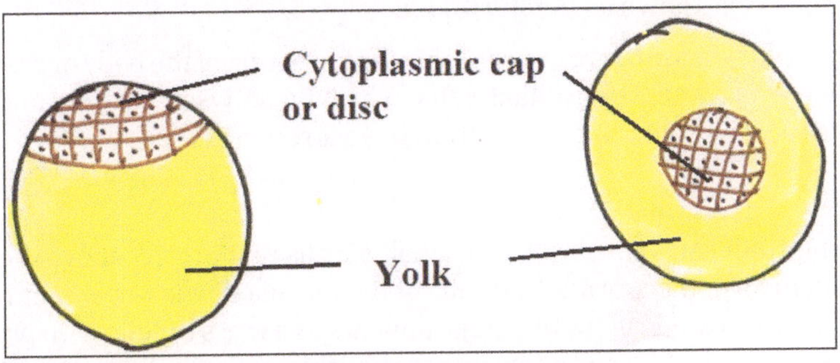

Meroblastic Discoidal Cleavage.

Teleost egg is a macrolecithal and highly telolecithal egg.

Cleavage

Cleavage is the process by which fertilized eggs undergo divisions to form blastomeres.

Teleost egg exhibits meroblastic discoidal cleavage.

The segmentation starts 30 minutes after fertilisation. The first cleavage is meridional, passing from the animal to vegetal pole. It is followed by cleavage in a number of vertical planes. Further horizontal cleavage divides the blastomeres into several rows.

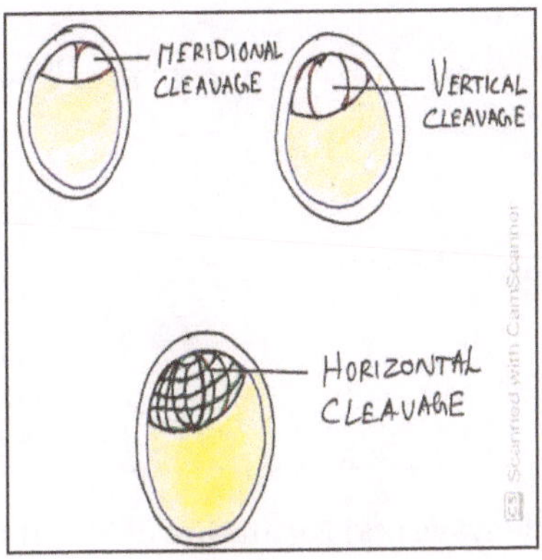

Cleavage.

Cleavage occurs only in the cytoplasmic cap on the animal pole; the yolk and the layer of cytoplasm surrounding it remain uncleaved. Even the cytoplasmic cap of the animal is not completely subdivided into cells. The deeper parts of the cytoplasm are provided with nuclei in the process of cleavage, but the cytoplasmic mass fails to divide and remains in the form of a syncytial layer adjoining the uncleaved yolk.

The disc of cells thus formed on the animal pole of the egg is known as the **blastoderm**. The syncytial layer of cytoplasm above the uncleaved yolk and below the blastoderm is known as the **Periblast.**

Only the blastoderm is responsible for the formation of the body of the embryo, while the periblast or **trophoblast cells**, which lie between the yolk and cells of the blastoderm helps in the mobilization of yolk reserves.

Morula

The blastoderm is converted into a ball of cells that project above the surface of the yolk to form the morula. The cells of the morula divide further and become smaller in size. In a side view, this stage appears as a mass of cells with prominent hemispherical projection and a convex base which rests in a hollow concavity of the yolk. The central cells of the blastoderm divide to form a number of free blastomeres, which becomes arranged as a syncytial layer on top of the yolk.

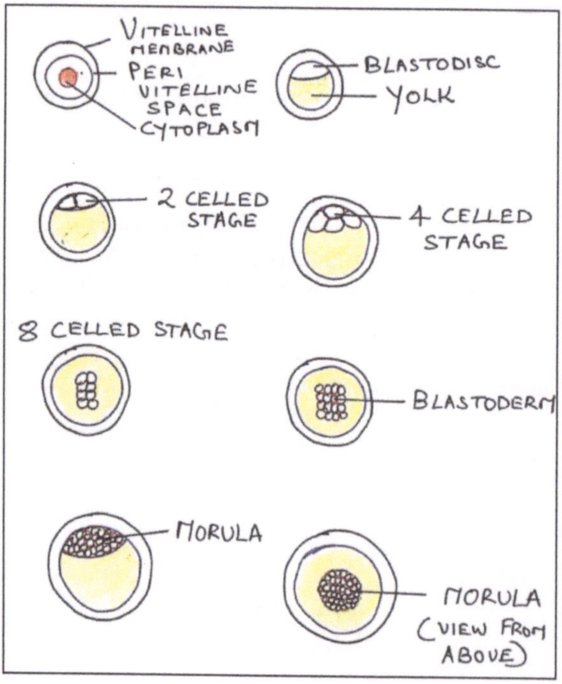

Cleavage and Formation of Morula.

Blastula

The hemispherical mass of cells of the morula project from the yolk. The cells then get flattened and extend outwards. The periphery of the blastodisc lines with the periphery of the yolk. The marginal or peripheral cells remain in close contact with the periblast, whereas the central cells of the floor of the blastodisc are raised. A cavity appears between the blastoderm and periblast by the elevation and splitting off of the floor of the central blastoderm from the underlying periblast. The cavity is the blastocoel. The blastoderm gives rise to the embryo and the periblast or trophoblast cells, which lie between the yolk and blastoderm is syncytial in nature and serves to digest the yolk and supply it to the developing embryo.

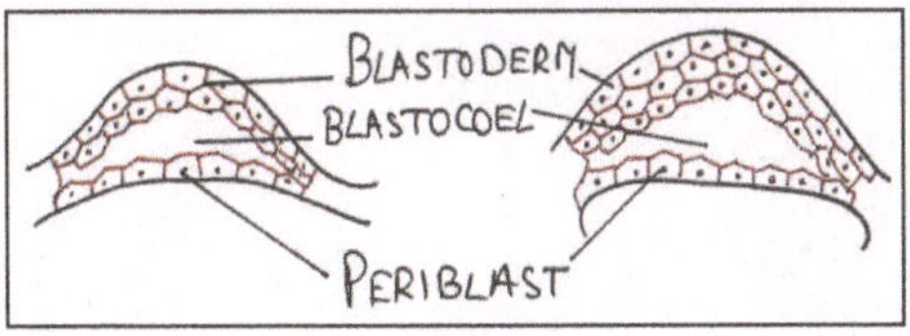

Blastula.

At the end of segmentation, the blastodisc becomes radially symmetrical. The radial symmetry changes to bilateral symmetry, because one region of the blastodisc becomes thicker. The future embryo develops from the thick sector and its median plane becomes the middle region of the embryo. The distal part of the thicker sector is the prospective posterior part of the embryo and its central part correspond to the prospective anterior region of the embryo.

Fate Map of Blastula

During blastulation, the cells get arranged and organised into well-defined areas, that are destined to give rise to specific organs of the embryo. Fate map is a chart showing the fate of each part of an early embryo, the blastula

The cells in the anterior region of the blastoderm are the prospective ectodermal cells, those cells lying in the middle forms the mesoderm, while the cells in the caudal region form the endoderm. The future notochordal cells occupy a median position. The neural area lies in front of the notochord as a rod of cells.

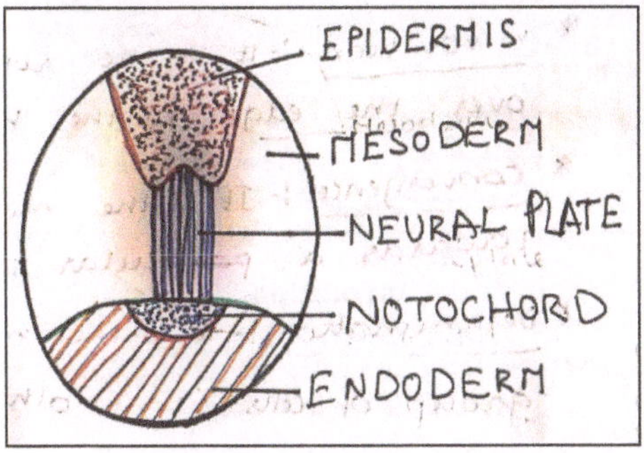

Fate Map of Blastula.

Gastrulation

Gastrulation is a dynamic process that involves morphogenetic movements of the prospective cells. The various regions of the blastula get reorganized to form a three layered structure which forms the body plan of the adult fish. Gastrulation occurs by involution, invagination, epiboly, convergence and delamination.

Invagination

It is the inpushing of a layer of cells to form a two layered structure.

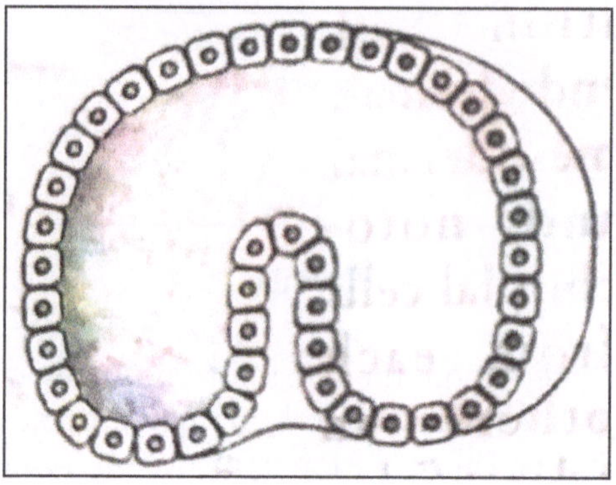

Invagination

Involution

It is the rolling in of cells over the edge of the blastopore lips.

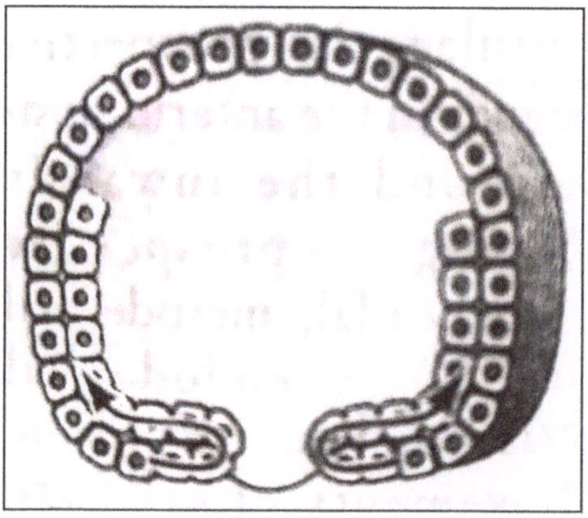

Involution.

Epiboly

It is the movement of ectodermal cells over the other areas.

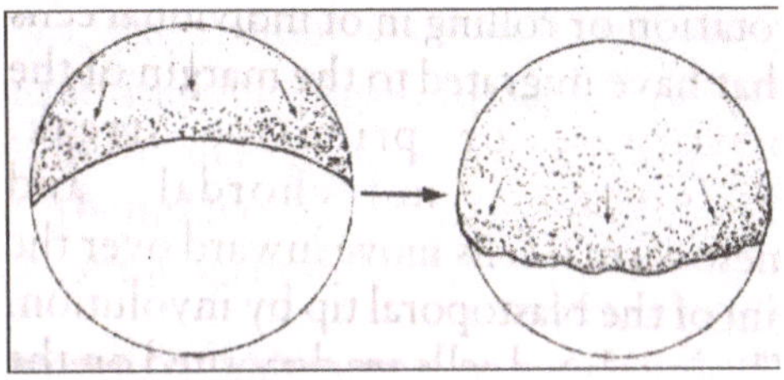

Convergence

It is the movement of cells towards a particular point.

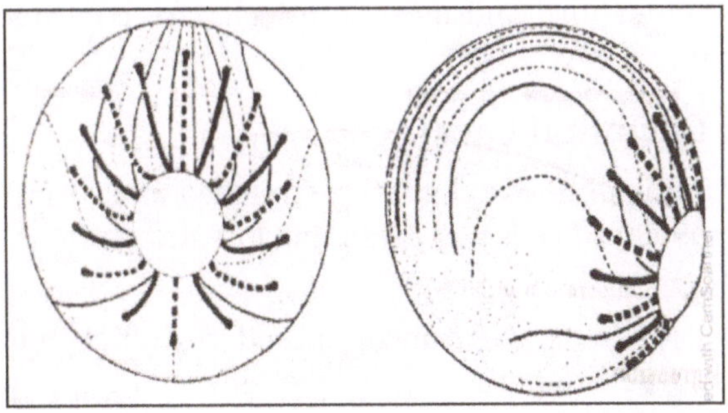

Epiboly

Delamination

It is the mass separation of a group of cells from other groups.

1. Invagination

Invagination takes place about 21-26 hours after fertilisation. The cells of the thicker sector invaginate at the limit of blastoderm and yolk, marking the beginning of gastrulation. The invagination extends around the edges of the entire blastoderm. The invaginated layer does not extend over the floor of the blastocoel; but is confined to the edges of the blastoderm, forming a prominent ring, known as the germ ring. The only part of the ring exhibiting further invagination is the thick sector of the blastoderm disc.

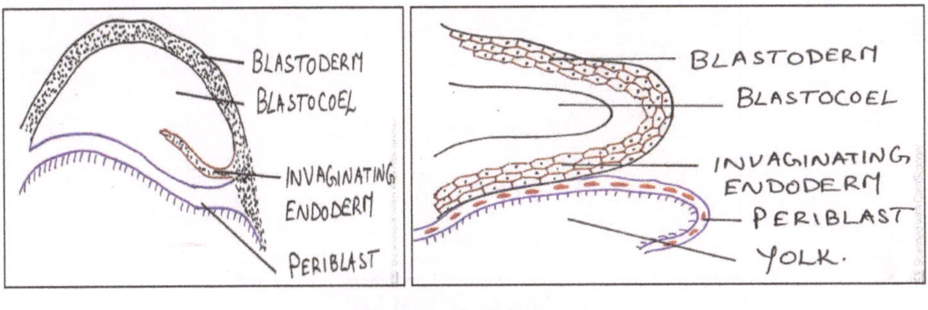

Invagination in Teleost Egg.

The prospective endodermal and mesodermal cells invaginate or turn inwards and migrate forward under the blastoderm, over the surface of the yolk to form a **hypoblast.**

2. Convergence

As the invaginating endodermal and mesodermal cells move forward, they converge or get concentrated towards the midline of the blastoderm.

3. Involution

The notochord cells pass inwards by rolling over the posterior edge of the blastoderm. The notochord and mesoderm cells are continuous forming the **chordamesoderm.**

Formation of Germinal Layers

The cells of the neural plate becomes stretched and forms a strip of tissue lying above the notochord cells and extending forwards from the posterior edge of the blastoderm.

Notochord cells get separated from the mesoderm by delamination. Mesoderm becomes thick and gets separated into segments lying on either side of the notochord. The endoderm becomes separated from the mesoderm and forms an endodermal plate in the middle line, below the mesoderm. The neural plate narrows and sinks deeper into the underlying tissue, separates from the epidermis and later becomes a part of the Central Nervous System.

The cells of the blastoderm continue to grow over the yolk by epiboly. The ectodermal cells grow over and cover the yolk mass from the outside forming a layer called the **epiblast**. The periblast grows and forms an inner covering of yolk. Both epiblast and periblast converge and close at the posterior end of the embryo to enclose the yolk, forming a yolk sac. Just before the yolk is completely closed in, a yolk plug may be seen projecting between the constricting edges of the blastoderm.

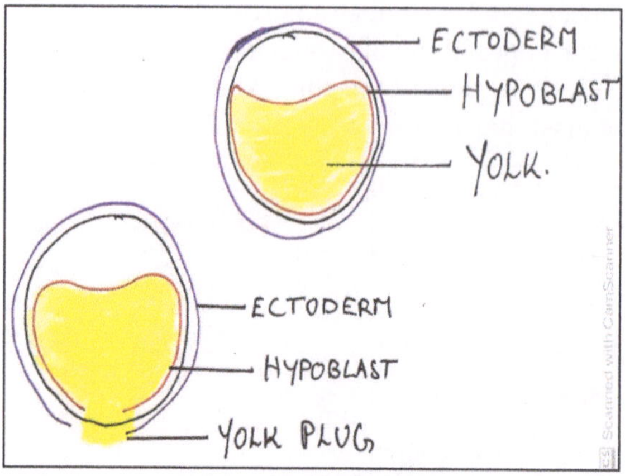

Epiboly in Teleost Egg.

Transitional Embryonic Development

On the completion of gastrulation, an embryo is formed, with a bilaterally symmetrical body, having a head, trunk and tail. A yolk sac is attached to the ventral side of the trunk by a yolk stalk. Yolk is digested by the periblast and supplied to the body through the blood vessels.

Various organs of the body are formed from the ectoderm, endoderm and mesoderm.

The ectoderm gives rise to the epidermis and its derivatives- enamel of the teeth, olfactory epithelium, lens of the eye *etc.* The neural plate gives rise to the brain, spinal cord and retina.

The Mesoderm divides into 3 parts – **Epimere**, **Mesomere** and **Hypomere.** The epimere gives rise to the appendicular skeleton, muscles, skin and scales.

The mesomere gives rise to kidneys, gonads, and their ducts. The hypomere gives rise to kidneys, gonads and their ducts.

The hypomere gives rise to the skeleton, coelom, heart, blood vessels, muscles of the head *etc.*

The endoderm gives rise to the germinal layers of gonads, inner epithelium of digestive glands and also contributes to the development of endocrine glands like thyroid, ultimobranchial glands *etc.*

Organogenesis

About 27-50 hrs after fertilisation, organogenesis begins.

1. Notogegenesis

The presumptive notochordal cells migrate inwards, thus forming a solid string like notochord.

2. Neurogenesis

The presumptive neural plate sinks down to the space vacated by the inwardly migrating notochord cells. The edges of the neural plate rise up and fuse with each other along the middle line, enclosing a cavity, the **neurocoel**. Thus a hollow tube like neural tube is formed, just above the notochord. The anterior part of neural tube swells to form the brain; whereas the part behind it forms the spinal cord. The optic lobes appear by lateral outgrowths of the forebrain.

Hatcing and Larval Development

As the embryo grows, the yolk sac shrinks in size. Hatching results from the softening of the chorion by the enzymes secreted by ectodermal glands in the pharynx.

The young ones at the time of hatching are called larvae or fry. The time required for larval development varies with the species. The larval development includes different stages.

Stage 1

- Light yellow in colour
- A row of pigments on the dorsal surface.
- Head is broad and the eyes are pigmented
- Yolk sac is large, broad towards the lower side and narrow towards the upper side.
- The median fins appear as a continuous fold.
- Pectoral fin is a small membranous flap without rays.

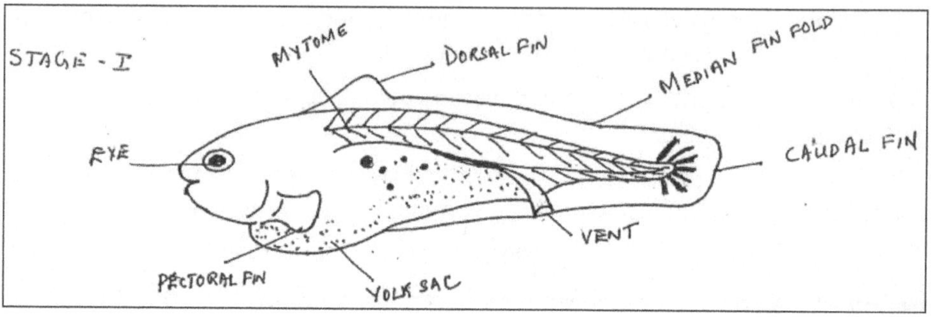

Stage 2

- Chromatophores are present in the head region
- Dorsal fin clearly demarcated with 7-8 rays
- Caudal fin with 15-16 rays.
- Anal fin visible, but without rays
- Air bladder becomes visible

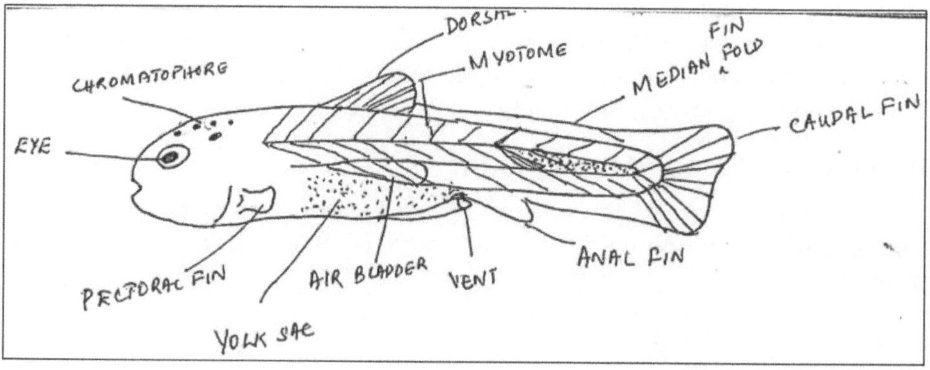

Stage 3

- Yolk sac is reduced in size
- Gape of mouth is large and extended backwards
- Dorsal fin is completely formed with large rays, but retains connection with the caudal fin
- Caudal fin is deeply forked with rays
- Anal fin with rudimentary rays.
- Pelvic fin visible as buds.
- Alimentary canal becomes visible.

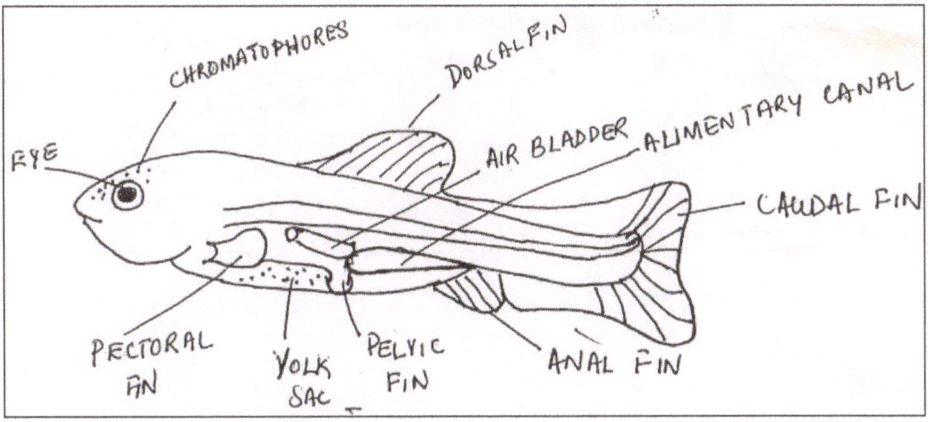

Stage 4

- The yolk sac becomes fully absorbed and is absent
- Dorsal fin is fully developed and is not connected with the caudal fin.
- Pelvic fins also developed with rays
- Pectoral fins without rays.

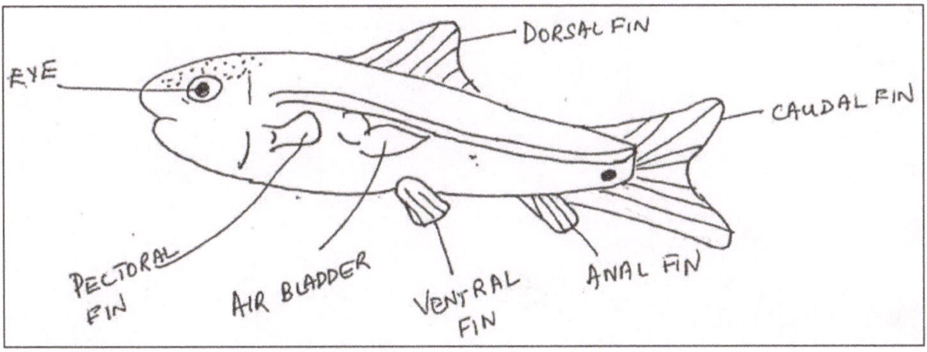

Stage 5

- The larva becomes a fingerling with adult characters
- All the fins are fully developed.
- Anal fins with rays and free from the caudal fins
- Pectoral fins fully developed with rays.
- Chromatophores present only on the dorsal side.

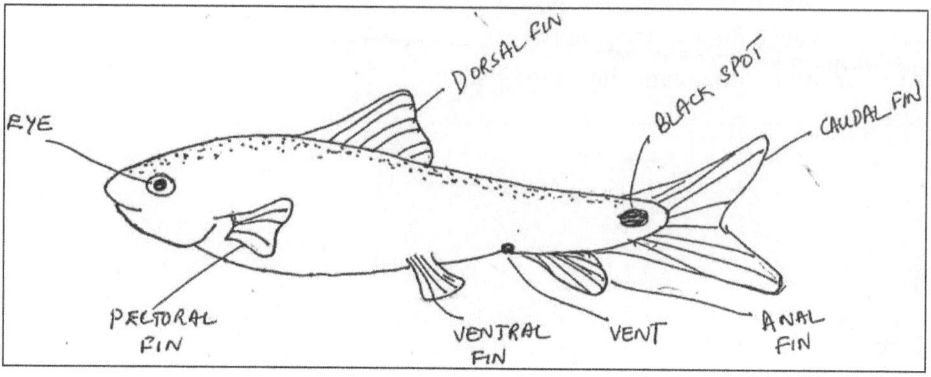

Chapter 10

Crustacean Neuroendocrine Systems

As in fishes, there are certain neurosecretory systems associated with the Central Nervous System of Crustaceans and producing important hormones involved in the physiology and reproductive behaviour. These neurosecretory systems are composed of modified nerve cells and associated neurohaemal organs. The important neuroendocrine systems in Crustaceans include:

1. X-Organ Sinus Gland complex
2. The Post Commissural organs (PCO)
3. The Pericardial organs (PO)

1. X-Organ Sinus Gland Complex

The nervous system of crustaceans include the supraoesophageal ganglion or brain, sub-oesophageal ganglion, circum esophageal connectives and a double ventral longitudinal nerve cord, with thoracic and abdominal ganglia.

From the brain, the optic nerve supplies the eye, travelling across the eye stalk. The optic nerve terminates in different optic centres in the eye stalk. The optic centres are:

a) Medulla terminalis, lying proximal to the brain
b) Medulla Interna
c) Medulla Externa
d) Lamina ganglionaris

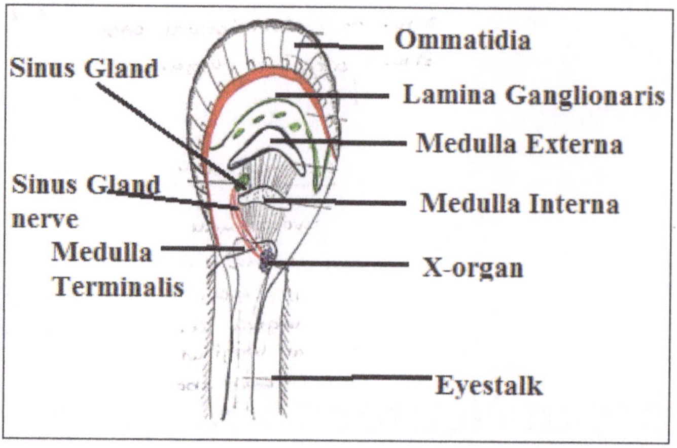

Sagittal Section of Crustacean Eyestalk.

Each optic centre of the optic nerve contains clusters of neurosecretory cells.

The Sinus Gland is a thick disc shaped, bluish white structure, lying on the surface of the optic ganglia on the dorsolateral side, at the level of medulla interna and medulla externa. The sinus gland is made up of bulbous nerve endings around a large blood containing space. A membrane, 1- 1.25 microns (μ) in thickness forms a boundary between the nerve endings of the sinus gland and the circulating haemolymph. The sinus gland lies under a layer of spongy connective tissue and under the muscles associated with the movement of the eyestalk.

The X-organ is a group of neurons on the distal ventral margin of the medulla terminalis. The axons of the neurosecretory cells of the X-organ form the sinus gland nerve. The axons of these cells converge to form a nerve fibre, which passes along the surface of medulla terminalis to end in the sinus gland. The secretions of the X-organ neurons pass along the nerve fibre to be released into the blood spaces of the sinus gland.

Hormones Produced by the Sinus Gland

The Sinus gland produces several hormones, which play an important role in the physiology of crustaceans. The hormones secreted by the X-organ Sinus gland complex are

1. Pigment Concentrating Hormone (PCH)

It is also called "Dark adapting hormone". It affects the concentration of red, white and black pigment granules, within the epidermal and retinal cells.

2. Pigment Dipersing Hormone (PDH)

It is also called "Light adapting Hormone" and controls the migration of retinal pigments in response to light. It is also known to affect the epidermal chromatophores.

3. Crustacean Hyperglycemic Hormone (CHH)

It influences the level of haemolymph sugar, primarily glucose. Its target organs are the hepatopancreas and abdominal muscles. CHH release is activated during moulting or stress, which increases the blood sugar level.

4. Gonad Inhibiting Hormone (GIH)

It is found to control the process of vitellogenesis in crustaceans and thus prevent gonad maturation. GIH is neither sex nor species specific and is a thermostable molecule. GIH inhibits binding of vitellin (yolk protein) to the microvilli of the oocyte, thus blocking vitellogenesis and gonad maturation.

5. Moult Inhibiting Hormone (MIH)

It inhibits the moulting process in crustaceans by preventing the secretion of moulting hormones by the Y- organs.

2. Post Commissural Organs (PCO)

The post commissural organs constitute another important neurohaemal structure in crustaceans. They are found posterior to the oesophagus and attached to the circumoesophageal connectives. They appear as two small bluish white swellings on the post - commissural nerves, which extend from the circumoesophageal commissure to a pair of muscles, associated with the exoskeleton.

Axons from the neurosecretory cells in the supraoesophageal ganglion or brain passes down the circumoesophageal connectives into the circumoesophageal commissure and bifurcate, sending one branch each to the post commissural nerves. The axons, on reaching the post commissural organs split into branches and end on an epineurium (blood space tissue) enclosing the dorsal blood sinus. The secretory axons terminating in the post commissural organs are from the cells within the brain and circumoesophageal connective ganglia. The post commissural organs lie near the point of attachment of certain muscles which attach the endophragmal skeleton to the dorsal carapace.

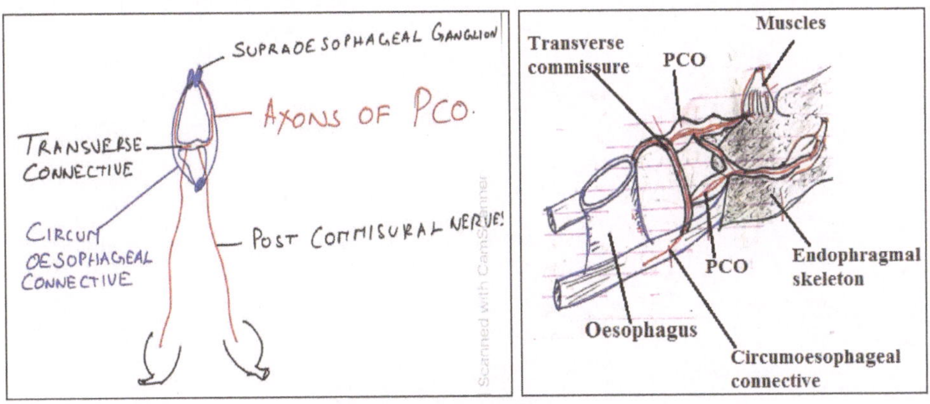

Post Commissural Organs.

(**Apodemes:** Any cuticular ingrowth of the body wall or in fold of exoskeleton for attachment of muscles.

Endophragmal skeleton: A complex internal skeletal structure formed by the fusion of apodemes, in decapods, providing a framework for attachment of muscles, generally not strongly calcified.)

Functions: The only hormonal activity associated with the Post - Commissural organs is the production of a factor causing pigment concentration in white chromatophores.

3. PERICARDIAL ORGANS (PO)

The crustacean pericardial organs are neurohaemal organs that lie in the pericardial or venous cavity surrounding the crustacean heart. The pericardial organs consist of groups of nerve endings attached to the lining of the pericardial cavity. They are located near the opening of large veins that enter the pericardial sacs.

The neurons constituting the pericardial organs have their cell bodies within the ganglia of the ventral nerve cord and suboesophageal ganglion.

The pericardial organs also include the **supporting** or **glial cells** of the peripheral region of the organ. Pericardial organs are the largest endocrine organs in crustacea.

Functions: The pericardial organs are neurohaemal structures that store and release factors that influence the heart contraction rate. The peptides that have been isolated from pericardial organs include the neuropeptide' **Proctolin** and the **Crustacean Cardioactive Peptite** (CCAP).

Chapter 11

True Endocrine Organs in Crustacea

In addition to the neuroendocrine organs, the crustaceans possess certain endocrine organs, composed of secretory tissues, which secrete the hormones. These hormones also play an important role in the physiology of crustaceans.

The true endocrine glands differ from the neuroendocrine system in that, they are not composed of modified nerve cells or neurons. The endocrine organs in crustaceans include

1. Y-organs
2. Androgenic Gland
3. Mandibular Organs
4. Gonads

1. Y-Organs

The Y-organs are endocrine glands, which influence the crustacean moulting process. They are derived from the ectoderm

Position

The Y-organs are located in the antennary or maxillary segments, positioned just beneath the ectoderm.

In crabs, the Y-organs are located at the anterio-lateral edge of the branchial chamber

Structure

Y-organs are small, compact and paired structures, only a few millimetres in diameter. In crabs, they are conical in shape, lenticular (lens like) in prawns and foliaceous (leaf like), in isopods and amphipods. The cells of the Y-organs are segregated into compact lobules. The cells are irregular in shape with large nuclei.

Functions

Y- organs are the source of moulting hormones. Moulting hormones are a group of steroids called **ecdysones**. In crustaceans, the moulting hormones constitute **Crustecdysone** (20 hydroxy ecdysone) and **Deoxycrustecdysone.** Moulting hormones are responsible for initiating moulting in crustaceans and also influence colour change, regeneration and gonad maturation. Moulting hormones are also necessary for the initial stages of spermatogenesis and oogenesis.

The Y- organs are under the control of moult inhibiting hormones produced by the sinus gland.

2. Androgenic Gland

It is an endocrine tissue situated on the genital tract, usually found attached to the distal region of the vas deferens. It lies alone on the sub terminal ejaculatory portion of the sperm duct, among the coxopodite muscles of the last walking leg, as a pyramidal mass of cells.

The androgenic gland is made up of a solid strand of cells folded back on itself several times. The length and width of the strand varies and the cells are arranged to form simple strands or lobules.

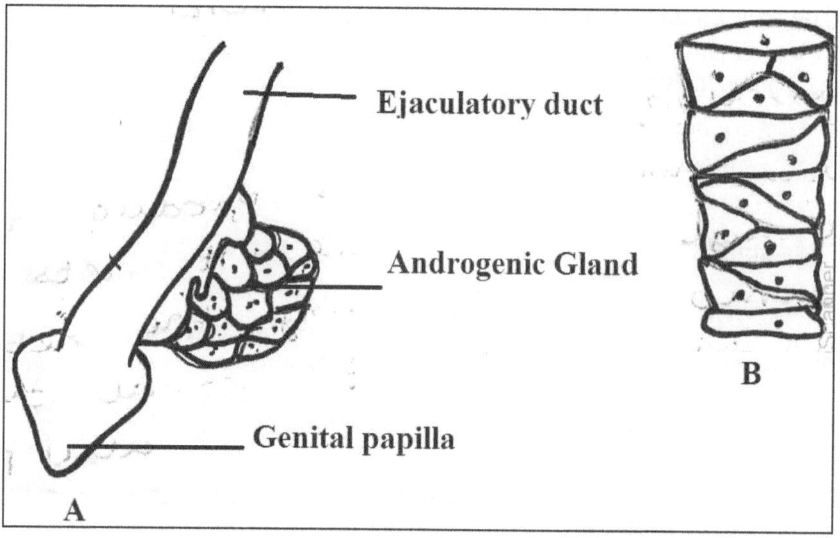

(A) Androgenic Gland (B) Cellular Strand of Androgenic Gland.

Functions

The androgenic gland produces a hormone – the **androgenic hormone**, which is involved in

- Development of primary and secondary sexual characters, in male and maintenance of maturity.
- Normal spermatogenesis.

3. Mandibular Organs

They are endocrine tissues lying close to the Y-organs. The cells constituting the mandibular organs are larger than that of the Y-organs and in some species, they are arranged in lobules and is more compact than the Y-organs. The cells contain numerous mitochondria and extensive rough endoplasmic reticulum associated with steroid secreting cells.

Functions

Mandibular organs produce a hormone called **Methyl farnesoate**. It acts like a juvenile hormone, promoting the retention or maintenance of larval characters. Mandibular organs also play an important role in reproduction, by regulating the female sexual cycle.

4. Gonads

The ovaries of amphipods and isopods are the source of one or more hormones that control the differentiation of the female secondary sexual characters. There is no direct evidence for the production of sex hormones by the ovaries in decapod crustaceans (crabs, Lobsters, Prawns).

REFERENCES

1. Subramoniam.T. (1993). *Spermatophores and Sperm Transfer in Marine Crustaceans.* In Blaxter.J.H.S. (Eds). Advance in Marine Biology. Vol 29. Academic Press.
3. Tombers. A.S. (1970). *An introduction to Invertebrate endocrinology.* Academic Press.
4. Chondar.S.L. (1980). *Hypophysation of Indian major carps.* Satish Books Enterprise.
5. Harvey.B.J. and Hoar. W.S. (1979). *Induced Breeding in Fish: Theory and Practice.* International Development Research Centre. Canada.
6. Indian National Science Academy. (1978). Symposium on hormonal steroids in fish. New Delhi.
7. Muir.J.F. and Ronald. J.Roberts. (Eds). (1993). *Recent advances in Aquaculture.* **Vol. IV**. Blackwell Scientific Publications.
8. Jamieson.B.G.M. (1991). Fish *Evolution and Systematics. Evidence from Spermatozoa.* Cambridge University Press.

Chapter 19- Principles of Biological cryopreservation.

Chapter 20- Live preservation of fish gametes.

9. Carl.E.Bond. (1996). *The biology of fishes*. (Edn.2) Saunders College Pub. USA.
10. Highnam.K.C.and Leonard Hill. (1977). The comparative endocrinology of Invertebrates. (Edn.2.). Elsevier.
11. Khanna.S.S. (2011). *An introduction to Fishes*. Silver Line Publications.
12. Kotpal.R.L. (2012). Modern Textbook of Zoology. Invertebrates, and Vertebrates. Rastogi Publications.
13. Lagler.K.F., Bardach.J.E., Miller. Robert.R and Dora.R. May Passino. (1977). *Ichthyology*. Wiley.
14. Matty.A.J. (1985).Fish endocrinology. Springer.
15. National Bureau of Fish Genetic Resources. (1986). Genetic improvement of fish stock and resource conservation. Bulletin No.1. NBFGR.
16. Varghese.T.J., Basavaraja.N, Nandeesha.M.C., Kesavanath,P., and Shetty.H.P.C. (1991). *Use of hormones for sex manipulation and growth promotion in cultivable fishes.* In Sinha.V.R.P. and Srivastava.H.C. (Eds). Aquaculture Productivity. Oxford and IBH Publishing Company.
17. Talbot. H. Waterman (1960) (Eds). *The Physiology of Crustacea*. Academic Press.

    Vol II. Sense organs, Integration and Behaviour.

    Vol. I. Metabolism and growth
18. Grizzle, John.M. (2204) *Reproductive Biology* (*Chapter 6.*) In Tucker.C.S. and Hargreaves.J.A. (Eds). (2004).Biology and Culture of Channel catfish. Elsevier Publications.
19. Turner, Daniel.C. and Bagnara, Joseph.T. (1971). Edn.5. General Endocrinology. W.B.Saunders and Company.
20. Hoar.W.S. and Randall.D.J. (Eds) Fish Physiology. Academic Press.

    (1969). Vol 2. The Endocrine System.

    (1969). Vol 3. Reproduction and Growth, Bioluminiscence, Pigments and Poisons.

    (1983). Vol. 9. Part A. Reproduction, Endocrine tissues and hormones.

    (1983). Vol.9. Part B. Reproductive Behaviour and Fertility control.
21. Hoar.W.S. (1966). General and Comparative Physiology. Prentice-Hall
22. Yadav.B.N. (1995). Fish endocrinology. Daya Books.

Section 2

Aquaculture

Chapter 12

Culture of Live Feed Organisms: Present Status and Recent Advancements

Introduction

Live food organisms include phytoplankton and zooplankton grazed upon by economically important fishes. According to David (2003) live foods are able to swim in water column and are constantly available to fish and shellfish larvae are likely to stimulate larval feeding response. In an aquatic ecosystem, these live food organisms constitute the most valuable resource for aquaculture. Most of the fish and shellfish larvae in nature feed on small phytoplanktonic and zooplanktonic organisms.

Live food organisms contain all the nutrients such as essential proteins, lipids. They are able to swim in water column and are available to fish and shellfish larvae thereby stimulate the feeding response. Micro algae have an important role in aquaculture as a means of enriching zooplankton for feeding fish and other larvae. In addition to providing protein and energy they provide other key nutrients such as vitamins, essential poly unsaturated fatty acids, pigments and sterols. Zooplankton forms the primary food for fish larvae. The dominant zooplankton groups are rotifers and copepods. These groups are the preferred prey for shrimp and fish larvae and are the most widely used live feeds by aquaculturists.

Artificial larval feeds are no match to live food organisms in terms of acceptance, nutritional and other factors. Feeding habit of fishes in natural water bodies is different among the species but all the fishes require protein rich live food for their better growth, efficient breeding and survival (Mandal *et al.,* 2009). Advances in live food enrichment technique have helped to boost the importance and potential of live food organisms in the raising of larval aquatic species.

The success in the hatchery production of fish fingerlings for stocking in the grow-out production system is largely dependent on the availability of suitable live food for feeding fish larvae, fry and fingerlings (Lim *et al.,* 2003). The availability of large quantities of live foods organisms such as marine rotifer, *Brachionus plicatilis* and *Brachionus rotundiformis* and *Artemia nauplii* to meet the different stages of fry production has contributed to the successful fry production of about 60 marine finfish species and 18 species of crustaceans. A common procedure during the culture of both larvae of fish and prawns is green water technology *i.e.* to add microalgae to intensive culture systems together with the zooplankton prey have become popular practice now a days (Tamaru *et al.,* 1994).

1. Microbes

Yeast is the common type among microbes and are an important ingredient in artificial larval diets. It can be directly used as a primary food source for many larvae but it is mainly used as a feed for zooplankton which is grown for use in larviculture. Yeast can also be used as supplement or replacement for algae in the feeding of post larval penaeid shrimps and also for bio enrichment of other live feed organisms.

Probiotics have been used as food supplement in fish culture. In contrast to the use in terrestrial animals, the use of bacteria as probiotics in case of fishes is no longer restricted to feeds only. Their use as a prophylactic measure is gaining momentum day by day in live food production to stabilize and promote larval rearing through a microbiologically balanced system. Common probiotics include commercial preparations of useful bacteria like *Bacillus subtilis, B. polyriyxa, B. negaterium etc. Lactobacillus acidophilus,* photosynthetic bacteria and yeast. These bacteria are consumed by several groups of zooplankton like protozoans, rotifers and copepods.

2. Micro-algae

Micro-algae are an essential food source in the rearing of all stages of larvae of several marine fish species, penaeid shrimps, marine bivalve molluscs like clams, oysters, and scallops, the larval stages of some marine gastropods like abalone, conch *etc.* and zooplankton. Intensive rearing of bivalves depends on the production of live algae, which comprises on average 30 per cent of the operating costs in a bivalve hatchery. The relative algal requirements of the various stages of the bivalve culture process depend on whether the operation aims at the mass-production of larvae for remote setting or growing millions of seed till planting size.

Algae are added during the non-feeding nauplius stage so that algae are available immediately upon molting into the protozoea stage. The most frequently used species in commercial mariculture operations are the diatoms *Skeletonema costatum, Thalassiosira pseudonana, Chaetoceros gracilis, C. calcitrans,* the flagellates *Isochrysis galbana, Tetraselmis suecica, Tetraselmis chui, Monochrysis lutheri* the chlorococcalean *Chlorella* spp. and *Nannochloropsis* spp. (Plates I-II.)

As feeding preference changes from primarily herbivorous to carnivorous during the mysis stages, the quantity of algae is reduced. Nevertheless, a background level of algae is maintained as this may stabilize water quality. Apart from the requirement for micro-algae for culturing and or enriching live prey organisms such as artemia and rotifers, algae are often used directly in the tanks. This "green water technique" is part of the commonly applied techniques for rearing larvae of marine fishes gilt head sea bream *Sparus aurata, Chanos chanos,* Mahimahi *Coryphaena hippurus,* halibut *Hippoglossus hippoglossus* and turbot *Scophthalmus maximus* as is the case with shrimps.

3. Yeast-based Diets

Because of their suitable particle size and high stability in the water column yeasts can easily be removed from suspension and ingested by filter-feeding organisms. More over yeasts can be mass-produced at a relatively low cost which is the main attraction as live feed. The potential of yeasts as a food in aquaculture has been proven by their successful application in the rearing of rotifers and some species of penaeid shrimp. However, a limited nutritional value of yeasts was reported for various species of filter feeders and attributed to their nutritionally deficient composition and/or undigestible cell wall.

4. Artemia

The brine shrimp *Artemia salina* is commonly, with a pan-global distribution in brackish habitats is one of the excellent live feed that are commonly used in marine hatcheries (Kumar and Babu, 2015). There are multiple strains within eight species (Hou *et al.,* 2006), although 90 per cent of the global trade in Artemia originates from the Great Salt Lake in Utah (Ruebhart *et al.,* 2008). Artemia can produce cysts that remain dormant for long periods if dehydrated. However, hatching is easily initiated through hydration, exposure to light and heavy aeration for 24 hours (Bengtson *et al.,* 1991). The ability to produce millions of individual artemia on demand, without the infrastructure required to breed them has made them the single most common live food in the aquaculture industry (Plate XV).

4a. Newly Hatched Artemia

Artemia nauplii, known as newly hatched artemia (Kumar and Babu, 2015) are the stage most commonly fed in the culture of marine ornamentals. Newly hatched artemia differ in size depending on species. They are typically between 400– 500 µm (Conceição *et al.,* 2010). Their size often limits their use as a first feed for most species of marine ornamental fish, as larval gape would have to be at least 800 µm. The lined seahorse fed newly hatched artemia displayed a significantly lower survival than conspecifics fed a nutritionally complete commercial diet (Vite-Garcia *et al.*, 2014). Newly hatched artemia are generally employed after larval gape size has increased. But reduced survival is often experienced and could be linked to the fact that nauplii may represent a nutritionally incomplete diet for larvae.

4b. Enriched Artemia

Enrichment of artemia can only be undertaken when the nauplii are at the second developmental phase, Instar II, reached between 27 and 32 hours after cyst hydration. Feeding begins at Instar II, and the nauplii can take on the diets nutritional profile (Ferreira de Sa, 2016). The benefit of using enriched artemia is that enrichment products are bio encapsulated in the nauplii (Sorgeloos *et al.*, 2001), which provide the larvae with the desired nutrient composition in order to attain fast growth rate.

Although enriched artemia are not commonly used as a first feed in the culture of marine ornamentals, they are useful in the initial rearing of ornamental species without a larval phase, such as seahorses. The direct development to a juvenile phase allows the consumption of larger first food items such as enriched artemia, as seen in the successful culture of various seahorse species. Although seahorses can be weaned onto dead prey successfully, enriched artemia is essential until individuals can tolerate the rapid water currents needed to keep non-living items suspended.

5. Zooplankton

5a. Rotifers

The Rotifera commonly called as "wheel bearers" are a group of tiny animals first observed by early microscopists in the late 1600s. They are relatively small group of minute, unsegmented, pseudocoelomate, aquatic invertebrates with bilateral symmetry. Around 2,000 rotifer species have been described. Most species are in the range of 0.1 to 1 mm in length, although a few species may reach 2 to 3 mm. Different rotifer species display a striking variety of body forms and the morphology of individuals may be further altered in response to ecological cues indicating the presence in the environment of particular types of prey or predators (Wallace and Snell, 1991).

Their size ranges from 90 to 350 µm. Body shape round and flat without spines. Rotifers serve as "living capsule" providing the nutrients required by the cultured marine fish larvae for proper development. Research on *B. plicatilis* and *B. rotundiformis* has increased enormously during the past three decades and these two species are the best-studied rotifers so far. They will probably maintain their role as food organism for fish larvae, in spite of attempts to replace them with more accessible formulated food.

Rotifers are small metazoans extensively used in aquaculture (Lee *et al.*, 2017) which, like artemia, are popular due to the ability to maintain high densities of

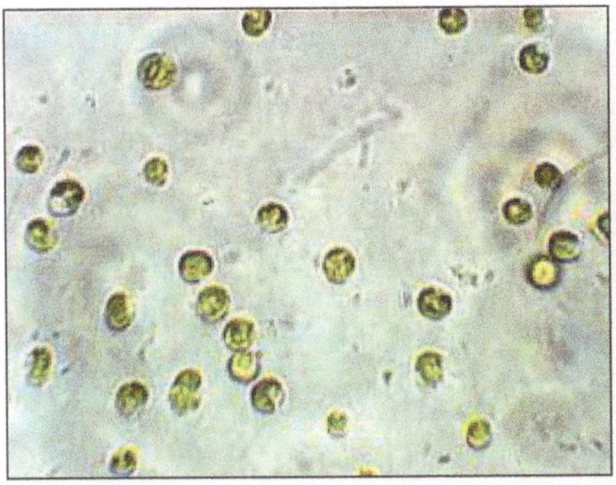

(I) *Isochrysis galbana*

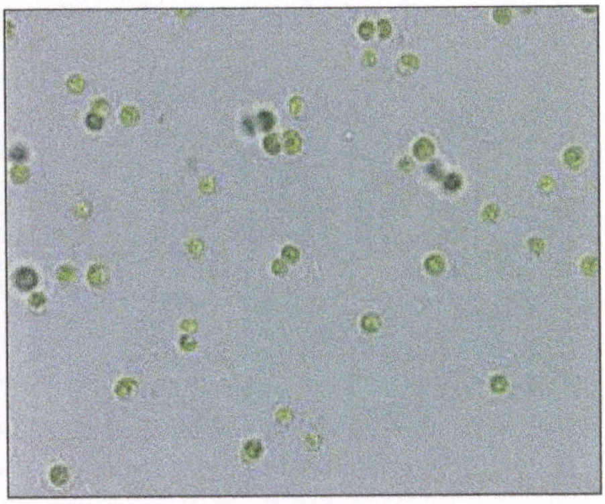

(II) *Chlorella vulgaris*

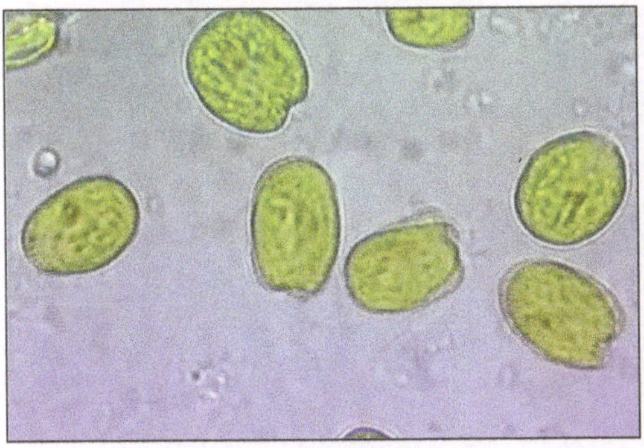

(III) *Tetraselimis chui*

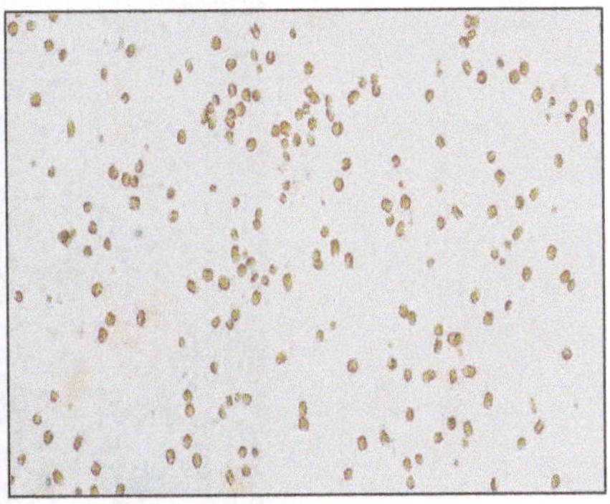

(IV) *Chaetoceros gracilis*

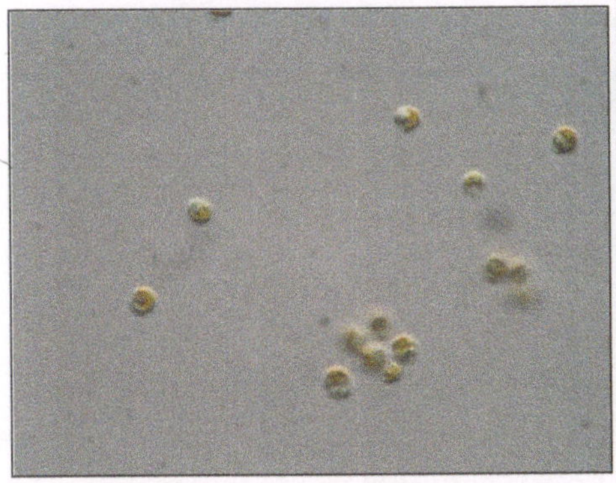

(V) *Nannochloropsis granulata*

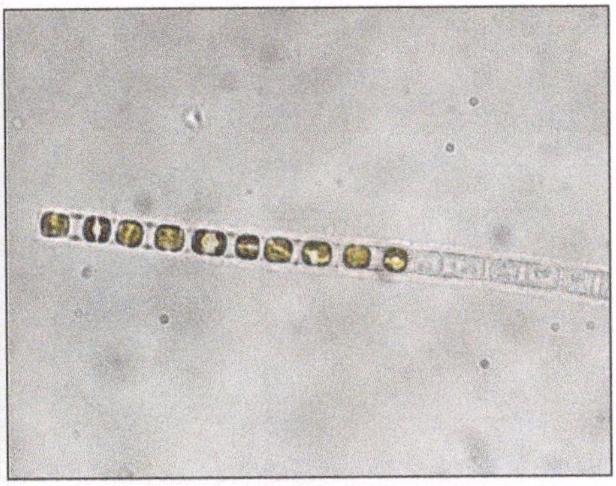

(VI) *Skeletonema costatum*

Plates I-VI: Live Feed Organisms.

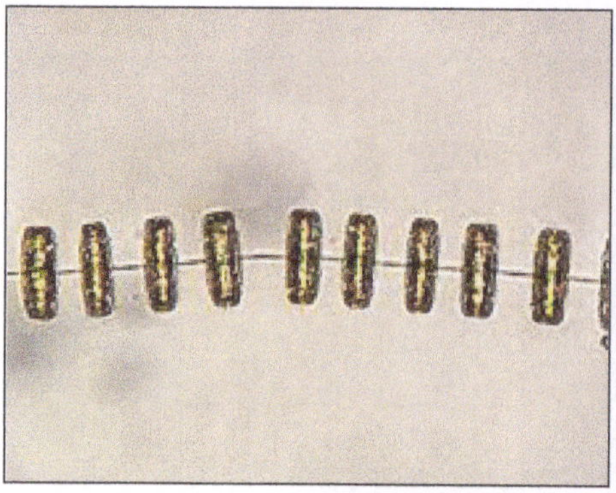

(VII) ***Thalassiosira psuedona***

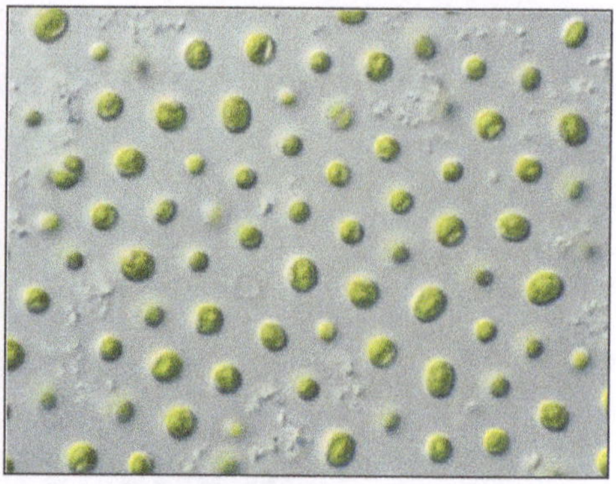

(VIII) ***Chaetoceros calcitrans***

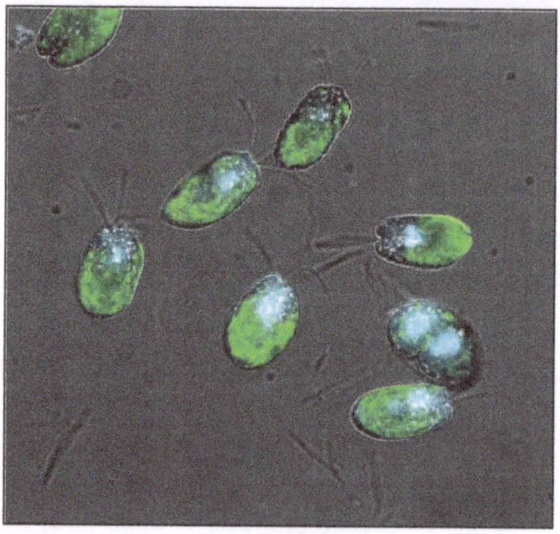

(IX) ***Tetraselimis suecia***

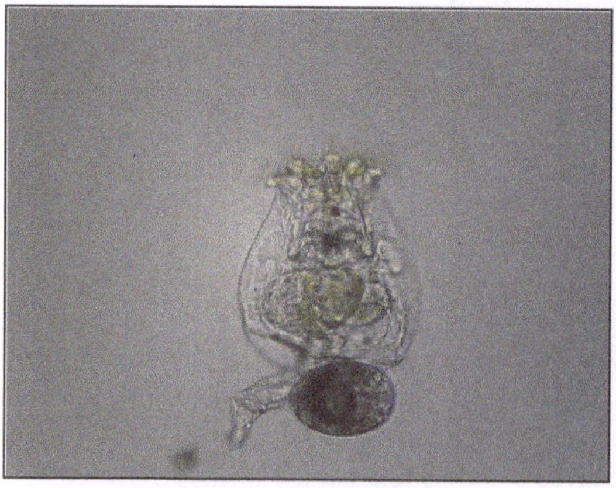

(X) ***Brachionus plicatilis***

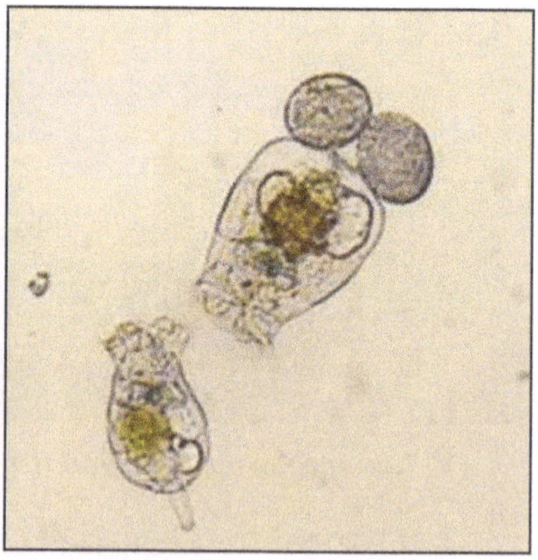

(XI) ***Brachionus rotundiformes***

Plates VII-XI: Live Feed Organisms.

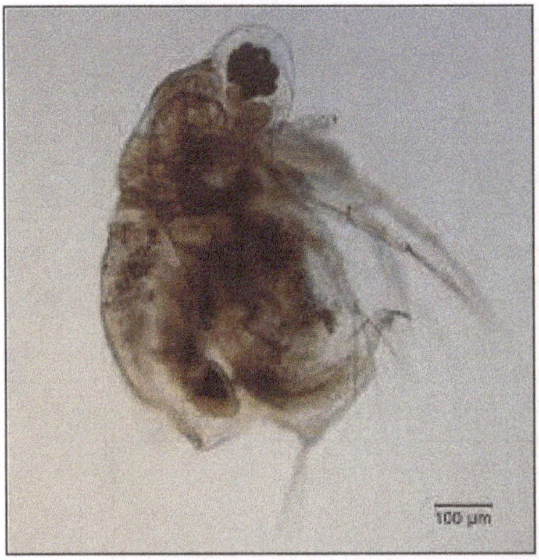

(XII) *Moina macrura*

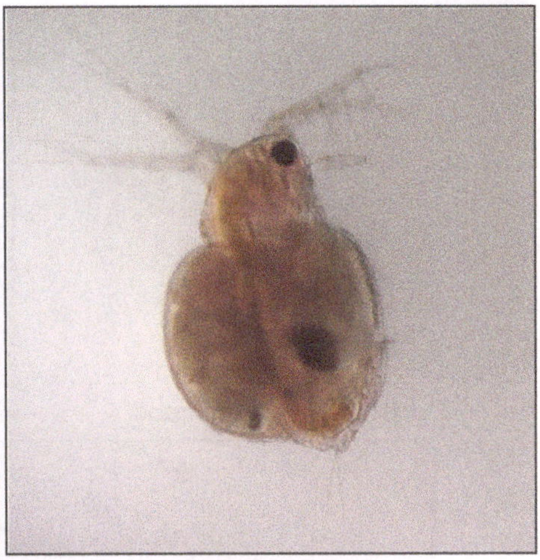

(XIII) *Moina macrocopa*

(XIV) *Daphnia magna*

(XV) *Artemia salina*

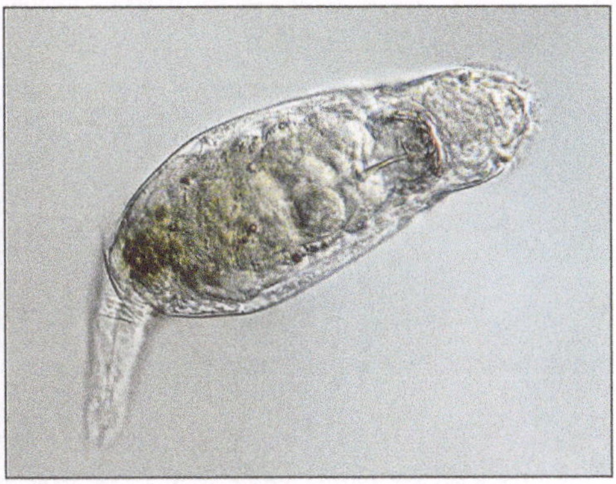

(XVI) *Proales similis*

Plates XII-XVI: Live Feed Organisms.

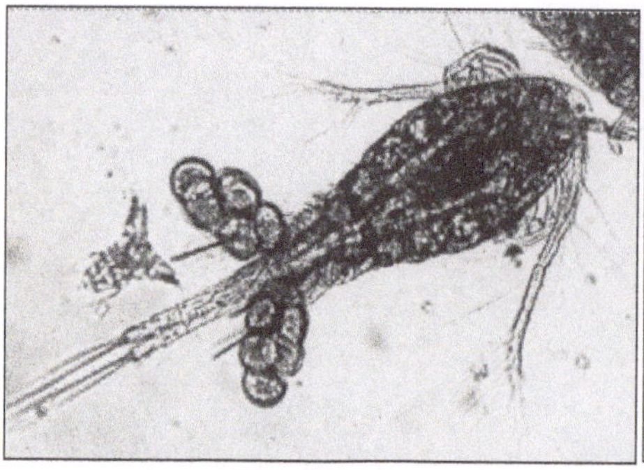

(XVII) Calanoid copepod

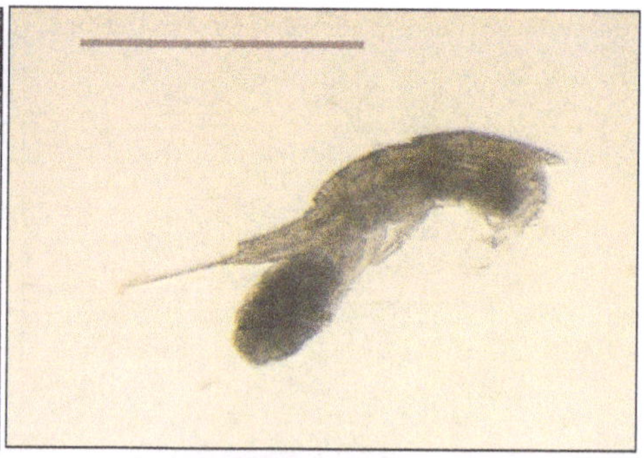

(XVIII) *Euterpina acutifrons*

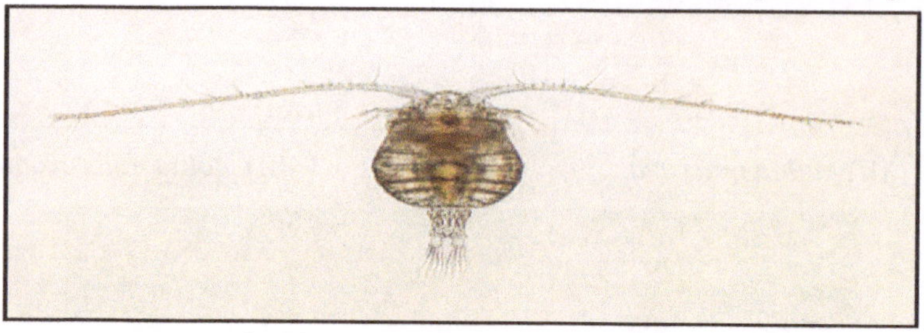

(XIX) *Parvocalanus crassirostris*

(XX) *Tubifex*

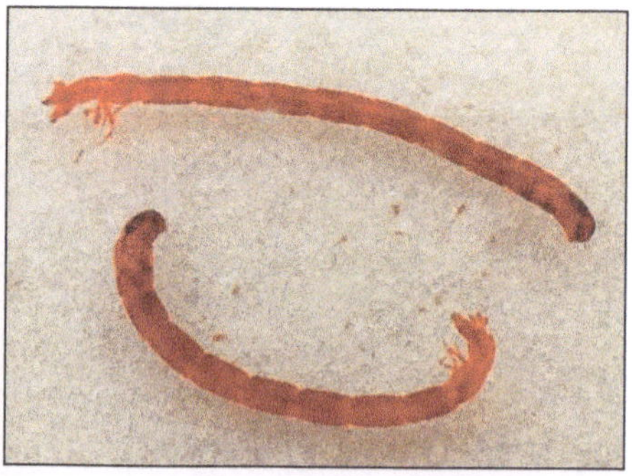

(XXI) Chironomid larvae

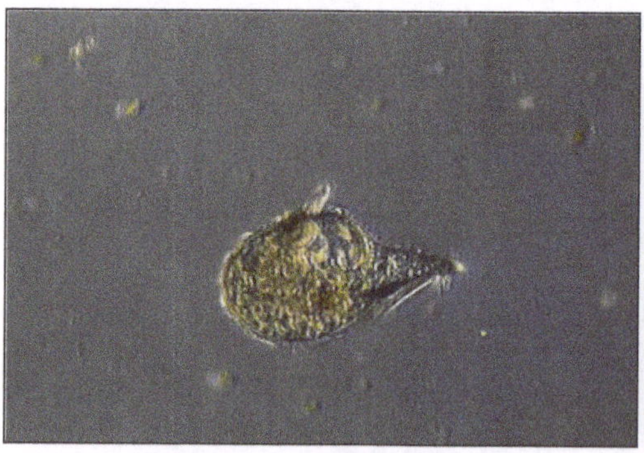

(XXII) *Fabrea salina*

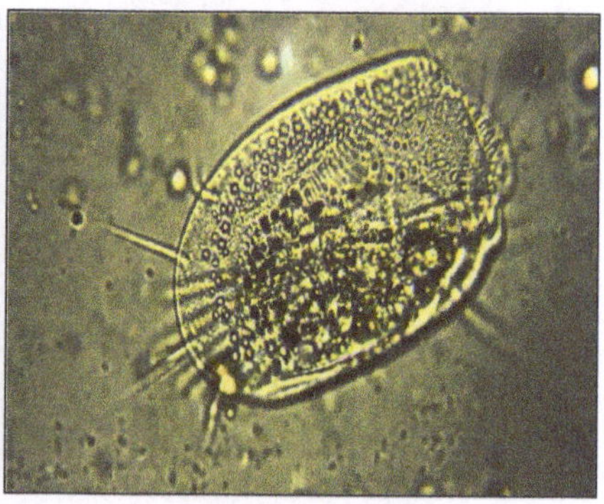

(XXIII) *Euplotes*

Plates XVII-XXIII: Live Feed Organisms.

individuals in a limited space. They are not hatched on demand but maintained in live cultures. By using advanced techniques, cultures can achieve 160,000 individuals.mL -1 (Yoshimura *et al.,* 2003). In 1995 it was found that *B. plicatilis* is more likely a species complex, with *Brachionus rotundiformis* being newly identified (Dhont *et al.,* 2013). Consequently, aquaculture uses a method of differentiation based on three sizes, large, small and super small of rotifer. (Plate X, XI)

The rotifer *Brachionus rotundiformis* can be mass cultivated in large quantities and is an important live feed in aquaculture. It is commonly offered to larvae during the first 7–30 days of exogenous feeding. Variation in prey density affects larval fish feeding rates, rations, activity, evacuation time, growth rates and growth efficiencies. *B. rotundiformis* can be supplied at the food concentrations required for meeting larval metabolic demands and yielding high survival rates. Live food may enhance the digestive processes of larval predators. Larvae are first fed on a small strain of rotifers, and as they increase in size, a larger strain of rotifers is introduced.

Rotifers have been used as live feed for cultured marine fish, since four decades. It is well known that a continuous, stable and reliable supply of nutritionally adequate rotifers is the key to the larviculture of marine finfish. Rotifers of the species *Brachionus rotundiformis* and *B. plicatilis* are almost indispensable for larval rearing of most marine finfish. The success of rotifer cultivation is dependent on selecting the most suitable rotifer species or strain for local culture conditions, maintaining water quality in culture tanks and choosing the most appropriate culture technique. Size, the type of reproduction and reproductive rates are species or strain specific. Culture temperatures, salinities, type of food and its quality are the factors that influence the type of reproduction and its rates.

Mass production of rotifers is achieved by encouraging rotifers to reproduce asexually, since sexual reproduction results in males and resting eggs. The amount of food that has to be supplied daily to each tank depends on the reproductive rate of rotifers. Usually 1-4 g of baker's yeast is supplied per million rotifers per day. The optimal range of pH for culturing rotifers is 7.5 – 8.5 and the pH affects the percentage of unionized ammonia in the water. The pH of cultures play an important role since the toxicity of ammonia is a function of pH, temperature and salinity.

5b. Copepods

Copepods are the most important group of zooplankton which forms the natural food for many fishes and invertebrates. The advantages of copepods over rotifers are that copepods have wide range of body sizes both within and between species. The early stage nauplii and copepodites can be extremely useful as initial prey for species that have very small larvae with small mouth gape at first feeding. Mostly species belonging to the orders Calanoida, Cyclopoida and Harpacticoida are popularly used as live feed in hatcheries. Species belonging to the genera, *Acartia, Pseudocalanus, Centropages, Eurytemora, Euterpina, Tigriopus, Calanus, Temora, Paracalanus, Pseudodiaptomus, Tisbe, Oithona and Apocyclopus* are widely cultured for hatchery use (Stottrup, 2000, 2003, 2006).

They are rich in PUFA, DHA and EPA, in most desirable ratios easily digestible (Stottrup, 2000) and rich in antioxidants, astaxanthine, vitamin C, D and E Copepods if fed during the larval phase, reduces malpigmentation and deformity rates, increases the pigmentation and survival (Hamre *et al.,* 2005).

Copepods are successfully cultured in fin fish hatcheries of many countries including India, especially for feeding atresial larvae of certain fishes like groupers. Many copepod nauplii are less than 100μ in size. Alone or as a supplement, in many cases copepods showed to improved primary growth than rotifers and brine shrimps. The smaller size of copepods enables their feeding by mouth gap-limited fish larvae like that of groupers and snappers.

i. *Proales similis*

Proales similis, a euryhaline rotifer with a typical body size of approximately 83 μm long and 40 μm wide is a relatively new addition to aquaculture feeds after being discovered in 2004 by Hagiwara *et al.* (2014). Due to its small size, *P. similis* has potential for rearing marine fish larvae which require first food smaller than *B. rotundiformis*. *Proales similis* have been used successfully to raise *C. ferrugata,* a species from a family that have initial gape sizes of approximately 160 μm (Hagiwara *et al.,* 2014). Despite the positive result from a highly prized ornamental family (Baensch, 2017), the wider application of *P. similis* appears under researched and work with higher value marine ornamentals, including species of dwarf angelfish, should be undertaken to explore the use of this species to the industry (Plate XVI).

ii. Calanoid Copepods

Copepods are the most abundant animals in ocean environments. Their nauplii are the natural prey of most wild fish larvae making copepod nauplii a more appropriate diet than rotifers or artemia nauplii for culturing fish. Copepod nauplii are an attractive prey item to larval fish as their erratic "zigzag" motion provides a visual stimulant to foraging animals. Copepod nauplii are also considered excellent live food since no enrichment is required, due to the high levels of fatty acids already provided by their diet. Even when fed a monoalgal diet of *Isochrysis galbana,* the calanoid species *Pseudodiaptomus hessei* attained DHA/EPA ratio of 2:1 (Siqwepu *et al.,* 2017) which is promising in larviculture of fin fishes. Copepods offer other nutritional benefits beside fatty acid provision, containing 700 times more iodine than artemia. This helps in the production of thyroid hormone, helping to regulate metamorphosis in fish (Alajmi and Zeng, 2013) (Plate XVII).

iii. *Euterpina acutifrons*

E. acutifrons is an atypical harpacticoid (Camus and Zeng, 2012). Gopakumar and Santhosh (2009), stated that adults were associated with the bottom of the tank, but the nauplii were present in the water column, a characteristic of some "true" harpacticoid species. There also appears to be some debate over the size of this copepod. Freshly hatched nauplii are recorded at 50–60 μm long and 40–45 μm wide (Gopakumar and Santhosh, 2009). Larval gape was measured at 293–437 μm and first feed size calculated at between 68–170 μm. *Euterpina acutifrons* nauplii were found in the gut at three days post hatch indicating its suitability as a first feed for this commercially valuable species (Leu *et al.,* 2015) (Plate XVIII).

iv. *Parvocalanus crassirostris*

They are small pelagic calanoid copepods with size measuring 100-400 micron that have been identified as a good candidate for culturing many species of food fish and ornamental fish. At 50-100 microns, the nauplii of copepods are a perfect size for small marine fish larvae.The calanoid copepod *P. crassirostris* has small nauplii, with recorded sizes of 62 μm long and 38 μm wide and 68 μm long and 56 μm wide making this species potentially suitable for larvae with the smallest gape sizes (Alajmi, *et al.,* 2013). *Parvocalanus crassirostris* nauplii use led to the first successful culture of the flame angelfish. Larvae consumed copepod nauplii between 60–70 μm at first feeding (Baensch, 2017). It is suggested that *P. crassirostris* may be suitable for the mass rearing of many other dwarf angelfish species, which are the most heavily traded ornamental angelfish (Baensch, 2017). *Parvocalanus crassirostris* nauplii were used to similar effect in the first confirmed captive metamorphosis of *P. hepatus*. *Parvocalanus crassirostris* nauplii were offered from days 3–11, with larger prey items introduced afterwards (DiMaggio *et al.,* 2017). Feeding experiments with this species proved that it is a crucial live feed for very small-mouthed larvae prior to transitioning onto larger feeds (Plate XIV).

5c. Cladocerans

They are commonly called 'water fleas'. Cladocerans belong to the sub-class branchiopoda and class crustacea of the phylum arthropoda. Two cladocerans, namely daphnia and moina are important as live food. Daphnia is found in freshwater ponds, tanks and lakes, all over the world. It swims by rapid jerks of the two large antennules and contains a broad spectrum of digestive enzymes such as proteases, peptidases, amylase, lipase and even cellulase which serve as exoenzymes in the gut of fish and prawns (Plate XIV).

Being larger in size than moina, it serves as live food for advanced stages of fishes. *Moina* are primarily inhabitants of temporary ponds or ditches. It is smaller in size ranging from 0.5 to 2 mm than daphnia containing 70 per cent more protein and therefore, goes well as a replacement for artemia in aqua hatcheries. Moina has also been extensively utilized as live food in many hatcheries and in the maintenance and culture of aquarium fishes of commercial importance (Martin *et al.,* 2006). In ornamental fish culture, Moina used to be the most common live food organism for feeding young fish larvae (Plate XII, XIII)

The nutritional content of Moina varies considerably depending on their age and the type of food they are receiving. The protein content of Moina usually averages 50 per cent of the dry weight. Adults normally have a higher fat content than juveniles. The total amount of fat per dry weight is 20-27 per cent for adult females and 4-6 percent for juveniles.

6. Tubifex

Tubifex is a type of worm of class - Oligochaeta under the phylum - Annelida. Clusters of these worms are commonly seen in sewage drains and in inshore areas and coastal muddy habitats. When disturbed, they jerk into the mud. Though not for larval and post-larval stages, but Tubifex make an ideally suited diet for brooders of various ornamental fishes. Among the natural food organisms, red worm, Tubifex *tubifex* is one of the best candidates owing its short generation time, occurrence in a vast range of habitats and tolerance to a wide spectrum of environmental variables (Kaster, 1980) (Plate XX).

7. Chironomid Larvae

Chironomids are one of the most ubiquitous, diverse, and ecologically important groups of aquatic macro invertebrates (Coffman and Ferrington, 1984). Chironomid is an insect belonging to the order diptera of the class insecta of phylum arthropoda. These are commonly called as blood worms due to the presence of haemoglobin in their body fluid (Madlen, 2005). Larvae initially live in soft tubes made from organic matter. Fully grown larvae are dark red in colour.

The chironomid larvae are recognized as an important food for many fishes and cultured invertebrate.These live foods are very popular in aquarium fish trade. Chironomid larvae are excellent source of protein, lipid, vitamins and minerals. The relatively high protein content (56 per cent), the high digestibility (73.6 per cent) and the apparent function in small quantities as a growth promoter in fish and crustacean diets make chironomid larvae a rich food for many organisms (Plate XXI).

8. Nematodes

The common species *Panagrellus revivus* and *Caenorhabditis elegans* can be cultivated very cheaply under laboratory conditions, especially the soil inhabiting species. Both these species can be cultured on a medium of breakfast cereal, soya peptone, yeast extract and liver extract mixed with sterile water. The nematodes have a short generation time ranging from 5-7 days and a high level of reproduction. Nematodes can be used for bioencapsulation. Experiments done with the antibacterial Romet-30 have produced good results. Enrichment is simply done by adding the product to the culture medium as direct enrichment or by placing the nematodes in an emulsion of the product as secondary enrichment.

9. Marine Infusoria

a. *Fabrea salina*

It is a floating pelagic form found in stagnant marine water conditions and is highly nutritious and delicious food for early stages of fish and crustacean. Size ranges from 40-60 microns. It reproduces in wild at temperature of 5-40°C and salinity up to 100 ppt. It has been found that a temperature of 37-40°C and salinity of 45-60 ppt is suitable for culture under controlled conditions. The young ones reproduce after 20-22 hrs. It can also be cultured using yeast, fermented wheat flour, fermented rice extract. A maximum density of 100 individuals/ml can be obtained in 5-7 days (Plate XXII).

b. *Euplotes* sp.

It is a saline benthic tiny organism found in stagnant water where there is accumulation of decaying organic matter. It feeds on bacteria and can be used to feed early stages of fish. It can be cultured in controlled conditions to obtain a density of 15000 individuals/ml after 5-6 days. Optimum temperature between 20-25 °C and salinity of 20-35 ppt is required for its culture (Plate XXIII).

10. Preserved Algae

A possible alternative to on-site algal culture could be the preservation of algae that are produced at relatively low cost in a large facility under optimal climatological conditions and using the most cost-effective production systems. Centrifugation of algae into a paste form and subsequent refrigeration until required is widely applied in North America by oyster hatcheries using remote setting techniques. The limited shelf-life and/the high prices of the products are the main disadvantages with preserved algae or algal pastes.

11. Micro-encapsulated Diets

Bioencapsulation or bio enrichment is the process involved in improving the nutritional status of live food organisms either by feeding or incorporating within them, various kinds of nutrients. Through micro-encapsulation techniques dietary ingredients can be encapsulated within digestible capsules and delivered to suspension-feeders without losses of nutrients to the aqueous medium. Common problems arising from the use of micro particulate feeds include settling, clumping and bacterial degradation of the particles, leaching of nutrients, and low digestibility of the cell wall material. In this regard, low susceptibility to bacterial attack and high digestibility for the filter-feeder may be the challenges in this regard.

Enrichment of Live Food Organisms

The most important factor governing the nutritional quality of live feeds for aquaculture practices is the essential fatty acid content, particularly eicosapentanoic acid and docosahexanoic acid, commonly called as highly unsaturated fatty acids.In recent years, much emphasis has been given to enhance the nutritional status of live food organisms through various techniques of enrichment and bioencapsulation. The nutrient which are lacking or present insufficiently in food organisms can be made available by allowing them to grow for a defined period in a medium containing appropriate quantities of the required nutrients.

Today, various kinds of micro particulate and emulsified formulations are used for boosting these live foods with essential fatty acids and other crucial components, including pigment and vitamins. The filter feeding nature of Artemia allows a very convenient way to manipulate biochemical composition. To improve the lipid composition of both Artemia nauplii and juveniles, researchers (Sorgeloos *et al.,* 2001) have developed several enrichment products like unicellular algae, compound diets, micro particulate diets, *etc.* The use of liposomes as enrichment products provides different advantages and possibilities (Tonheim *et al.,* 2000). This outstanding achievement in larviculture has resulted in increased survival, higher growth rate and greater resistance to stress and disease. The enrichment technique is becoming more prevalent for intensification of culture practices and in the future, it will help initiate commercial culture of new marine fish as well as ornamental fish species.

Major Constrains in Live Feed Culture

Live feed remains the most practical solution for larval rearing for marine as well as ornamental fish species. It is essential to maintain a steady supply of adequate quantities of live feed in intensive culture systems to meet the dietary requirements of several stages in the development of fin fishes, crustaceans and molluscs. The main constraints of live feed production are its cost of production, especially in smaller hatcheries. Difficulties in getting pure strain, lack of infrastructure facility like controlled environmental laboratory for culture maintenance *etc.* are other concerns in this area. Disease aspects should also be considered as live feed also act as carrier of diseases to the larvae of fish and shellfish. So maintenance of hygiene is very important during their production in hatcheries.

Enrichment process of various live feeds that are available for commercial production and diversification of feed items are to be given more emphasis. The new technology of enrichment process is a costly affair for poor and medium level farmers. Similarly, the high infrastructure and labour requirement along with the variable cost for live feed production illustrates the need to develop suitable modified culture technology. Among live feeds the more expensive one is artemia. More care should be given to selection of local strains of artemia as the import of cysts is a costly affair. Nutritional status of the live feed organisms needs to be properly illustrated for feeding different larval stages of fish and shellfish. More research thrust should be given on nutritional studies suitability of many of the available live food organisms.

Future Prospects

A good selection of micro algal species is available to support the aquaculture industry. However for some particular applications for industry sectors, new species with improved nutritional quality or growth characteristics could improve hatchery efficiency. The use of microalgae either as a full or partial enrichment should be considered for improving the nutritional quality of zooplankton. Finfish producers are concerned with improving the quality, quantity and cost effectiveness of their live feed production facilities. Many of them now supplement cultures with omega yeast, vitamin E, D, C and B12, marine oils or other HUFA sources, and vitamin B12 producing bacteria to improve feed quality. Today, live feeds for fish larvae are being improved by adjusting their biochemistry through controlling their diet and supplementing the cultures with microencapsulated feeds or emulsified oils. While algae and rotifers are the most widely used live food items, their use is not without problems and limitations. One of the main problems with the bio enrichment is that rotifer and copepod cultures are subject to collapse under mixed conditions.

Conclusion

The current preferred protocol for feeding marine ornamental fish larvae of most commercially cultured species, by transitioning from rotifers to artemia nauplii and then to enriched Artemia, has limited wider application to enable a shift in future aquaculture development. In order to diversify the number of species cultured within the marine ornamental fish trade, there is a need to increase the availability of a wider range of small sized live feeds. So more studies are needed in the area of selection of new species of algae and zooplankton for larviculture of marine species. Copepod nauplii are an excellent choice of live feed before transitioning to rotifers. However, the current limitations of their culture need to be overcome. Therefore, consideration should be given to developments necessary to overcome the bottlenecks currently being experienced within the commercial ornamental fish aquaculture trade. The most crucial area lacking data is the initial gape dimensions of larvae. It would be pertinent to record the first gape dimensions of any spawned species in a communal database to allow live foods to be appropriately sized to maximise the chances of successful rearing.

Further investigations should focus on the replacement, or at least supplementation, of both rotifer and artemia with copepod nauplii of varying sizes hatched from diapause eggs. This would be the most significant advancement in the culture of marine ornamental fish. In addition, significantly more work is required to develop approaches for creating and storing diapause eggs from copepods. In order to achieve this, further research is needed to determine which species produce diapause eggs. Progression in storage techniques is required to allow the integration of this food source into mainstream aquaculture and to diversify species cultured. The use of diapause copepod eggs would allow utilisation of a more effective live food, without the resources and risks associated with current feeds.

REFERENCES

Alajmi, F., Zeng, C. 2013. Evaluation of microalgal diets for the intensive cultivation of the tropical calanoid copepod, *Parvocalanus crassirostris*. Aquaculture Research 46: 1025–1038.

Baensch, F. 2017. Dwarf angelfish. In: Marine Ornamental Species Aquaculture, Calado, R., Olivotto, I., Oliver, M.P., Holt, J. G. (Eds.), Wiley Blackwell Publishing, Chichester, pp. 279–298.

Bengtson, D.A., Leger, P., Sorgeloos, P. 1991. Use of Artemia as a food source for aquaculture. In: Artemia biology, Browne, R.A., Sorgeloos, P., Trotman, C.N.A, (Eds.) CRC Press, Boca Raton, Florida, pp. 255– 285.

Camus, T., Zeng, C. 2012. Reproductive performance, survival and development of nauplii and copepodites, sex ratio and adult life expectancy of the harpacticoid copepod, Euterpina acutifrons, fed different microalgal diets. Aquaculture Research 43: 1159–1169.

Coffman, W. I., Ferrington, L. C., 1984. Chironomidae. In: Merritt, R. W and Cummins, K. W. (Eds.), An introduction to the aquatic insects of North America (2nd ed.). Kendall-Hunt Publishing Co., Dubuque, Iowa. pp. 551-652.

Conceição, L.E.C., Yufera, M., Makridis, P., Morais, S., Dinis, M.T. 2010. Live feeds for early stages of fish rearing. Aquaculture Research 41: 613–640.

David A. B., 2003. Status of marine aquaculture in relation to live prey: past, present and future. In: Josianne, G. S and Lesley, A. M. (Eds.), Live feeds in marine aquaculture. Blackwellpublishing, UK, pp. 1-16.

DiMaggio, M.A., Cassiano, E.J., Barden, K.P., Ramee, S.W., Ohs, C.L., Watson, C.A. 2017. First record of captive larval culture and metamorphosis of the pacific blue tang, *Paracanthurus hepatus.* Journal of the World Aquaculture Society 48: 393–401.

Dhont, J., Dierckens, K., Støttrup, J.G., Van Stappen, G., Wille, M., Sorgeloos, P. 2013. Rotifers, Artemia and copepods as live feeds for fish larvae in aquaculture. In: Advances in Aquaculture Hatchery Technology, Allan, G. Burnell, G. (Eds.), Wood head Publishing, Oxford, pp. 157–202.

Ferreira de Sa, T.L. 2016. Substitution of Instar I by enriched Instar II Artemia in the first days of *Solea senegalensis* rearing. Masters thesis, Institute of Biomedical Sciences Abel Salazar, University of Porto, Portugal. 93 pp. https: //repositorio-aberto.up.pt/handle/10216/89555 (Accessed 20 April 2020).

Gopakumar, G and Santhosh, I (2009) *Use of Copepods as Live Feed for Larviculture of Damselfishes.* Asian Fisheries Science, 22 (1). pp. 1-6.

Hagiwara, A., Wullur, S., Marcial, H.S., Narisato, H., Sakakura, Y. 2014. Euryhaline rotifer *Proales similis* as initial live food for rearing fish with small mouth. Aquaculture 432: 470–474

Hamre, K., Moren, M., Solbakkan, J., Opstad and I., Pittman, K. 2005. The impact of nutrition on metamorphosis in Atlantic halibut (*Hippoglossus hippoglossus* L.). Aquaculture 250: 555-565.

Hou, L., Bi, X., Zou, X., He, C., Yang, L., Qu, R., Liu, Z. 2006. Molecular systematics of bisexual Artemia populations. Aquaculture Research 37: 671–680.

Kaster, J. L., 1980. The reproductive biology of *Tubifex tubifex* Muller (Annelida: Tubificidae). American Midland Naturalist, 104: 364-366.

Kumar, R.G., Babu, D.E. 2015. Effect of light, temperature and salinity on the growth of Artemia. International Journal of Engineering Science Invention 4: 7–14.

Lee, I.S., Ohs, C.L., Broach, J.S., DiMaggio, M.A., Watson, C.A. 2018. Determining live prey preferences of larval ornamental marine fish utilizing fluorescent microspheres. Aquaculture 490: 125–135

Leu, M., Sune, Y., Meng, P. 2015. First results of larval rearing and development of the blue striped angelfish *Chaetodontoplus septentrionalis* (Temminck and Schlegel) from hatching through juvenile stage with notes on its potential for aquaculture. Aquaculture Research 46: 1087–1100.

Lim, C. L., Dhert, P., Soregloos, P., 2003. Recent developments in the application of live feeds in the freshwater ornamental fish culture. Aquaculture, 227: 319-331

Madlen, M. H., 2005. Culture of chironomid larvae (insecta- dipterachironomidae) under different feeding systems. Egyptian Journal of Aquatic Research, 31(2): 403-418.

Mandal, S. C., Das, P., Singh, S. K., Bhagabati, S. K., 2009. Feeding of aquarium fishes with natural and artificial foods: available options and future needs. Aqua International, 3: 20-23.

Martin, L., Arenal, A., Fajardo, J., Pimental, E., Hidalgo, L., Pacheco, M., Garcia, C., Santiesteban, D., 2003. Complete and partial replacement of artemia naupli by Moina micura during early post larval culture of white shrimp *Litopenaeus schmitti*. Aquacult. Nutr., 12(2): 89-96.

Ruebhart, D.R., Cock, *I.E.*, Shaw, G.R. 2008. Brine shrimp bioassay: Importance of correct taxonomic identification of Artemia (Anostraca) species. Environmental Toxicology 23: 555–560.

Sorgeloos, P., Dhert, P., Candreva, P., 2001. Use of the brine shrimp, Artemia spp., in marine fish larviculture. Aquaculture, 200: 147–159.

Stottrup, J. G. 2006. A Review on the status and progress in rearing copepods for marine Larviculture. Advantages and disadvantages among Calanoid, Harpacticoid and Cyclopoid copepods. In: L. Elizabeth Cruz Suarez, Denis Rique Marie, Mireya Tapia Salazar, Martha G. Nieto Lopez, David A Villarreal Cavazos, Ana C. Puello Cruz and Armando Garcia Ortega, (Eds.). Advances en Nutrition Acuicola VIII. VIII Symposium International de Nutrition Acuicola. 15-17 November.

Stottrup, J. G. and McEvoy, L. A. 2003. Live Feeds in Marine Aquaculture. Ames, Iowa, USA: Iowa State Press. pp. 336.

Stottrup, J.G. 2000. The elusive copepods: their production and suitability in marine aquaculture. Aquaculture Research 31: 703-711.

Tamaru, C. S., Murashige, R., Lee, C. S., 1994. The paradox of using background phytoplankton during the larval culture of striped mullet, *Mugil cephalus* L. Aquaculture, 119: 167-174.

Tonheim S. K., Koven, W., Ronnestad I., 2000. Enrichment of Artemia with free methionine. Aquaculture, 190: 223–235.

Vite-Garcia, N., Simoes, N., Arjona, O., Mascaro, M., Palacios, E. 2014. Growth and survival of *Hippocampus erectus* (Perry, 1810) juveniles fed on Artemia with different HUFA levels. Latin American Journal of Aquatic Research 42: 150–159.

Yoshimura, K., Tanaka, K., Yoshimatsu, T. 2003. A novel culture system for the ultra-high density production of the rotifer, *Brachionus rotundiformis*- a preliminary report. Aquaculture 227: 165–172.

Wallace, R. L. and Snell, T. W. 1991. Rotifera. In: Ecology and Classification of North American Freshwater Invertebrates (Ed. by J. H. Thorpe and A. P. Covich), Academic Press, New York. pp. 187 - 247.

Chapter 13

Zebrafish: A Promising Model in Scientific Research

Aswathy Sivaraman[1] *and Binu Ramachandran*[2]*

Smriti Brain Research Laboratory (SBRL),
[1]*Department of Zoology, Christ College (Autonomous), Irinjalakuda, University of Calicut, Thrissur, Kerala-680125, India*
**Laboratory of Neuronal Plasticity, Department of Zoology, University of Calicut, Thenhipalam, Kerala-673635, India*
**e-mail: binuramachandran@uoc.ac.in*

ABSTRACT

Animal models are undoubtedly non-human species used in the clinical translational research. [1]. There are several distinct types of animal models are currently used in the modern field of biomedical research. Mice, rat, fishes, monkeys, etc. are the commonly used animal models, and in them mice representing the excessively used animal models. Because of the physiological and anatomical similarities that mammalian models have with humans, they are a valuable human-disease model [2]. However, the high costs connected with their rearing and management make them a challenging model to associate with. Zebrafish (Danio rerio), a small cyprinid teleost fish introduced in the scientific field as early as 1930s as a developmental and embryological model organism. Zebrafish research progressed from developmental and embryological studies to genetic studies including cloning, mutagenesis, transgenics, etc. during 1980s. Through 1990s zebrafish became established as a model organism to deal with several human-diseases. Now, the scientific community is accepted the Zebrafish as an internationally accepted model organism and extensively using in the field of translation research by following the rule of 3Rs (Replacement, Reduction and Refinement) as required by the multiplicity of national and international regulatory bodies [3].

Taxonomic Position and Natural History of Zebrafish (*Danio rerio*)

The zebrafish was scientifically named *Danio rerio* in 1822 by Scottish physician Francis Hamilton, along with a few other species of the genus from Eastern India. Danio was divided into two subgenera in 1916: Danio and Brachydanio, the latter of which included zebrafish. Danio and Brachydanio were synonymized in 1991, and Danio was reclassified in 2003 into the genera Devario and Danio, which includes zebrafish. The zebrafish is one of more than 20 species in the Danio genus and one of over 4,000 species in the Cypriniformes order [4].

Table 13.1: Phylogenetic Classification of Zebrafish (*Danio rerio*). Classification below the gnathostome level is following Betancur study [5].

Classification
Kingdom : Metazoa (Animalia)
Super phylum : Deuterostomia
Phylum : Chordata
Subphylum : Vertebrata
Infraphylum : Gnathostomata
Megaclass : Osteichthyes (Euteleostome)
Superclass : Actinopterygii
Class : Actinoptergii
Subclass : Neopterygii
Infraclass : Teleostei
Super cohort : Clupeocephala
Cohort : Otomorpha
Subcohort : Ostariophysi
Superorder : Cypriniphysae
Order : Cypriniformes
Suborder : Cyprinoidae
Family : Danionidae
Genus : *Danio*
Species : *Danio rerio*

However, in strict evolutionary terms, all living vertebrates are fish, with tetrapod, such as ourselves, being a specialised type of (lobe-finned) fish capable of surviving on land. The human body and genome are littered with evidence of our own piscine evolutionary past. More than 500 million years ago, the chordate phylum had seen the emergence of the vertebrate subphylum [6]. The advent of the jaw mechanism was a crucial innovation of gnathostomes that facilitated the discovery of novel food sources, leading to jawed vertebrate's evolutionary success. The acquisition of two sets of paired fins, the pectoral and pelvic fins, was another key morphological innovation that led to the gnathostomes. Gnathostomes are separated further into two groups: cartilaginous fishes and bony vertebrates, which include zebrafish and humans. Around 400-450 million years ago, the lineages leading to zebrafish and humans split, with the ray-finned fishes (Actinopterygii), which include zebrafish, splitting from the lobe-finned fishes (Sarcopterygii), which gave rise to tetrapod. The zebrafish is a member of the teleost fishes (Teleostei), the most diverse group of ray-finned fishes. Teleost account for about half of all extant vertebrates, with over 25,000 species. Zebrafish was discovered from the Kosi River, West Bengal by Hamilton. As per the references *Danio rerio* was identified to distributed over rivers, small streams, channels and paddy field from Myanmmar, Pakistan, India, Bangladesh and Nepal [7].

The Zebrafish: Gender differences and Colour Pattern

Zebrafish is a sexually dimorphic teleost fish, but the gender differences are not readily apparent in the early stages of growth until it becomes sexually active. Male and female morphological distinctions in a sexually matured adult zebrafish include colour pattern, body size, and so on. In comparison to female zebrafish, male zebrafish have a fairly slim body form. Male zebrafish are substantially more energetic than females. On the ventral side of their bodies, they likewise appear to be more gold or yellow in colour than females (belly). Because of their bigger bellies, where eggs are deposited, females have far more distinguishing traits than males. Females move a little more slowly than males. A silvery-blue line on the lateral side of the body distinguishes females from males. An alternating pattern of four to five dark stripes and four light stripes covers the flank and the anal and tailfins of adult zebrafish [8].

Fishes contain multiple types of chromatophores involved in the production of colour pattern, unlike humans and birds, which have only one type of pigment cell (melanocytes). In zebrafish, CP generation requires interaction within and between chromatophore cell types dispersed in the hypodermis of the body and the epidermis of scales and fins. Chromatophores emerge directly from neural crest cells in the larval stage, but adult stripe patterns emerge during metamorphosis (3–6 weeks after fertilisation) and are composed of yellow xanthophores, silvery or blue iridophores, and black melanophores. Despite the fact that the anal and caudal fin stripe patterns are continuous with the body stripes, the method of fin stripe creation is different from that of the body stripes [9, 10].

The Zebrafish: Developmental Stages

External fertilisation occurs in zebrafish, where the embryo develops outside of the parent fish. The reliable monitoring of zebrafish early embryonic phases in experimental conditions is made possible by the transparent embryo growth. Zebrafish mating begins as soon as the light begins to appear. Rapid morphological transformation occurs during embryonic development, with the endpoint traditionally defined as hatching or birth. Zebrafish in the laboratory hatch 3-4 days after fertilisation (dpf). The protruding mouth stage (3 dpf at 28.5°C) marks the conclusion of embryogenesis. Stages of early development includes, a zygote (0 hpf), blastula (2 ¼ hpf), gastrula (5 ¼ hpf), pharyngula (24 hpf) and early larval stages (72hpf).

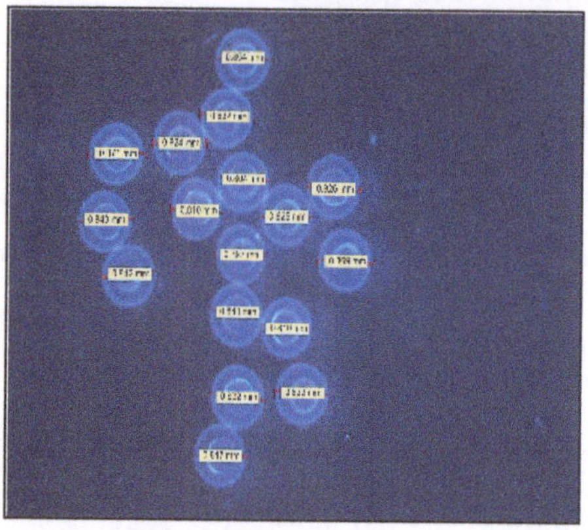

Figure 13.1: Zebrafish Embryo with Scaling.

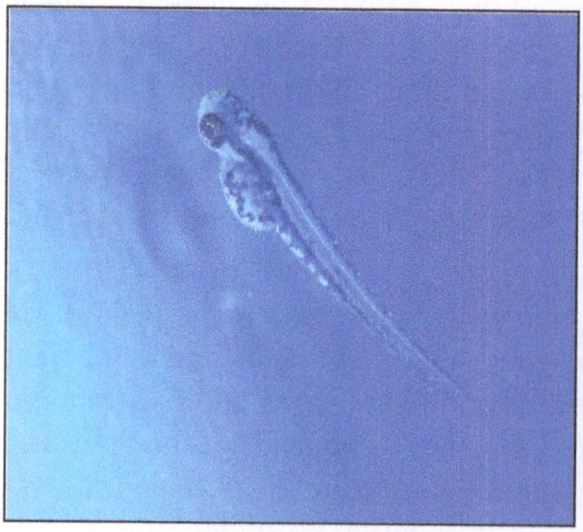

Figure 13.1: Zebrafish Free Hatching Larval Stage at 72hpf.

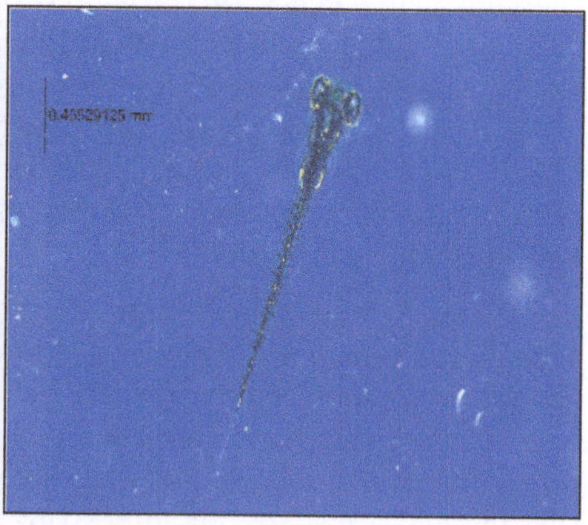

Figure 13.1Zebrafish of 96 hpf.

All the pictures are captured using leica stereo microscope.

Larval development lasts 6 weeks after the embryonic period, during which the fish more than triple in size and undergo a series of morphological changes that turn the fins, colour pattern, and general body morphology into the juvenile configuration. This occurs at 45 dpf at 28.5°C in zebrafish. Diet and microRNAs may have a role in the regulation of fish metamorphosis. Juveniles have lost their larval fin fold and gained scales, and their shape is similar to that of adults; nevertheless, they are not sexually mature. Zebrafish reach sexual maturity and are considered as adults after 3 months at 28.5°C [12].

Zebrafish are omnivores that ingest larval and adult insects, small crustaceans, and other zooplankton, as well as algae, plant material, and other detritus. As per the recent studies, intestinal bacteria are identical in recently caught wild zebrafish and domesticated lab strains, suggesting a core gut microbiota critical for growth and development [13].

Laboratory Maintenance of Zebrafish

1. Automated Aquatic Rack System for Zebrafish Rearing

Zebrafish are kept in a circulating system that filters and aerates the system water on a regular basis to ensure that the water quality required for a healthy aquatic environment is maintained. Excess food and fish excrement are also filtered by the circulation system. The room temperature or tank temperature is usually kept between 26-28.5 °C, and the illumination is set to 14:10 hr (light: dark). The pH of the system water should be monitored on a daily basis and kept between 6.8 and 7.5. To raise the pH, sodium bicarbonate should be administered when necessary. Recirculating aquatic rack systems, which consist of a parallel array of individual, isolated tanks, are widely used in the laboratory for rearing aquatic animals. Water is constantly recirculated in these systems between individual tanks and a central sump, passing through several filters along the way. Aquatic rack systems, for example, are extensively used to raise freshwater zebrafish and Xenopus frogs, [14,15] and widely used for commercial aquaculture[16,17]. Mechanical and biological filtration, as well as a proportion of daily water changes, are commonly used in these systems to eliminate accumulated waste materials. Before being pumped back into the tank, water from all of the tanks passes through a 120-micron filter pad, a 50-micron canister filter, a biological filter, an active carbon absorption filter, and a UV disinfection filter. In the zebrafish system, dechlorinated/ aged water is employed. Water that has been aged for at least 48 hours can be de-chlorinated. Water should be kept in a reservoir with a pump circulating it to keep it warm and speed up the de-chlorination process in ideal conditions. These types of systems are ideal for housing large groups of animals. The inclusion of numerous smaller tanks allows for the isolation of animals, which is especially important for sustaining huge numbers, various species, or distinct lines of animals (*e.g.*, inbred lines, mutants, or transgenic lines). This configuration also allows for controlled breeding mating.

Several individual polycarbonate tanks are stored on the rack. These tanks are secured by ridges moulded into their bottoms that engage the metal crossbars (shelves) of the aluminium rack, preventing them from moving. These tanks come in a variety of sizes and can be easily removed for access and cleaning. Cleaning fish tanks should be done on a regular basis. Close the water supply to the tank, drain excess water by tilting the tank backwards, and carefully remove the tank from the system. Dirt and algae growth will be visible on the tank's bottom and sides. Spray the fish net with 70 per cent ethanol, rinse it in water, and allow it to dry before reusing it. Remove the baffle from the filthy tank and saturate both sides with 70 per cent ethanol. Before reusing, thoroughly rinse with tap water and allow the tank and baffle to dry completely. To ensure proper operation, the circulating system filters must be checked and changed on a regular basis. These filters should be replaced on a regular basis to guarantee that all of the fish tanks receive adequate and clean water. Daily repositioning or replacement of the 120-micron filter pad is recommended; reposition towards the water flow to use it completely before replacing with a new one. The carbon filter should be replaced every two weeks (every two weeks). To replace the carbon filter, carefully remove the carbon filter unit with a wrench. Replace the activated carbon that has been used with new activated carbon. Replace the carbon holder in the filter unit and tighten it. Replace the filter unit in the system. Check that water is flowing to the filter after turning on the system. Before installing the filter, flush some water through the pipes to clean any dirt that has built up. Every six months, the biological filter should be cleaned. In the circulating system, a biological filter is usually placed between the canister and the carbon filter. Remove the filter unit from the system to change the biological filter. By pressing the pressure release button, the filter's pressure will be released. With a wrench, unscrew the filter. Remove the filter unit's lid. To separate the siporax from the water, empty the contents of the filter into a sieved container. UV filters are utilised to keep biological impurities (like bacteria) out of the system and should be updated every nine to ten months. It should be noted that the UV filter disinfection dosage rate is around 110 mJ/cm^2 at the start of the lamp's life and drops over time, thus the globe should be replaced even if it looks to be working [18].

Figure 13.4: Three Rack Automatic Filtration System for Zebrafish Housing.

2. Water Quality Parameters of Optimum Range [19]

Parameter	Range
Alkalinity	50-150 mg/L $CaCO_3$
pH	6.8-7.5 (6.0-8.5 tolerated)
Temperature	26-28.5 °C
Hardness	50-100 mg/L $CaCO_3$
Un-ionized Ammonia	6.0 mg/L
Salinity	0.5-1 g/L
Conductivity	300 -1,500 μS

3. Feeding

The food that fishes eat is crucial to their health and productivity. In captivity, fish may be fed live prey, artificial diets, or a combination of the two. It is possible to feed zebrafish a live food diet for the rest of their lives. This is particularly preferred since live feeds such as Artemia or rotifers (Brachionus sp.) often have balanced nutritional profiles and are thus more likely to suit the needs of zebrafish [20]. Bloodworms (Chironomid larvae), which are an essential part of the diet of wild zebrafish, are widely available and can be used to feed adults. Live feeds are also visually and chemically appealing to fish, as well as being highly digestible [21]. Feeding by body weight entails giving a daily diet based on a defined percentage of the fish's body weight. In intensive culture systems, larval fish are fed more often than adult fish, up to 50–300 per cent of their body weight per day, compared to 1–10 per cent for adults [22,23]. This strategy, which requires managers to have accurate estimations of total fish weight in the system, is widely used in commercial aquaculture, but it is rarely, if ever, used in research settings for zebrafish. However, it is likely the most efficient and scientifically valid method for determining zebrafish feeding allowances. It would necessitate the establishment of a set of "average" weights for larval, juvenile, and adult fish, which would be associated with age and overall length, as well as a standard number of fish per tank. The most optimal parameters for zebrafish can then be determined by feeding trials in which meals are delivered to fish at varied percentages of body weight.

4. Reproduction

Zebrafish are asynchronous batch spawners who, once sexual maturation is reached, spawn continuously under perfect conditions [24]. Females have the power to spawn almost every day. Females in continuous residence with a male spawned once every 1.9 days [25] and experiments showed that females in their studies could produce viable clutches every day for at least 12 days, albeit egg output varied significantly. This period is likely to be prolonged when the environment (water quality, diet, social context, *etc.*) is bad or the fish are used for production on a regular basis. Olfactory signals are vital in zebrafish reproduction. By releasing steroid glucuronides into the water, males encourage females to ovulate [26]. Following ovulation, females release hormones that promote male mating behaviour, which precedes and prompts oviposition and spawning. The ability of "dominant" female zebrafish to hold water and prevent inferior females from spawning has been established, demonstrating that pheromones have the ability to inhibit reproduction [27].When designing brood stock management regimes, this information should be taken into account; for example, combining increased chemical filtration with a faster rate of water replacement on recirculating systems may help to reduce potential decreases in brood stock reproductive output caused by pheromonally mediated dominance interactions between females. Zebrafish reproduction is heavily influenced by photoperiod. Zebrafish spawn in the laboratory [28] and in the wild at dawn, within the first few hours of sunlight. However, spawning does not appear to be limited to this time range. Zebrafish can reproduce at any time of day in captivity, especially in the evenings before a dark period is imposed, and viable eggs can be removed from females at any time of day. After a heavy downpour, zebrafish have been witnessed spawning in the wild in the afternoon. While there is a strong link between photoperiod, ovulation, and spawning, no complete control exists. There appears to be some sort of relationship selection in zebrafish. Females discriminate based on males' ritualised mating behaviour, as well as the building and defence of territory [29]. Females who are paired with certain males have been shown to produce larger clutches and spawn more frequently, supporting this notion [31]. On the other hand, the selective foundation of female choice is unknown. According to Spence and Smith, territorial males had somewhat better reproductive success than non territorial males at low densities, but there was no difference at greater densities, and male dominance rank did not correlate with female egg production. When it comes to picking males, females may use male territorial defence as a criterion. Both male and female zebrafish show a strong preference for oviposition site in both laboratory and field studies, preferring gravel over silt and spawning preferentially. In addition, fish prefer vegetated areas to non-vegetated habitats. As a result, it's possible that the zebrafish mating system is predicated on male defence of optimum spawning sites, which females are finicky about. Olfactory cues are employed by many fish, including zebrafish, to discriminate kin from non-kin, and this strategy could be used to prevent mating with close relatives during breeding. Zebrafish appear to employ smell signals in social and mating circumstances. Adult female zebrafish preferred the scents of non-related, unfamiliar (reared and maintained separately) males for mating above those of unfamiliar brothers, according to Gerlach and Lysiak's smell plume experiments [32]. Years of breeding zebrafish in captivity (often by sib-mating) have demonstrated that a female's preference for any or all of these features can be easily countered. On the other hand, a greater knowledge of reproductive behaviour may make it easier to develop better spawning procedures and breeding programmes, which could assist increase the efficiency of spawning in laboratory breeding facilities.

5. Raising of Larvae

Zebrafish larvae develop in 2.5–3 days post-fertilization (dpf) under standard laboratory settings (28.5 °C) [33]. After hatching, they use specialised cells on their heads to stick to hard surfaces [34] demonstrating low levels of spontaneous activity until 5 dpf, when they expand their gas bladders by consuming air at the water surface. Larvae subsist largely on yolk-sac reserves until the onset of exogenous feeding, which is coincident with completion of a number of physiological steps, including the development of a functional digestive system [35,36] and the ability to regulate position in the water column (gas bladder inflation. Depending on conditions, this transition takes place between 5– 6 dpf [11]The yolk is quickly exhausted after this point, with total absorption occurring <"7 dpf [37]. From this point on (and preferably before), feed must be provided on a nearly constant basis to meet nutritional demands, which are highest during this life stage [38].

Selection of the appropriate diet is essential to larval survival. Live prey items such as Paramecium, rotifers, and Artemia are preferable, due to the fact that they are generally more nutritionally balanced, visually and chemically attractive, digestible, and distributed evenly in the water column than artificial diets. Artificial diets may be utilized, but in general survival and growth rates are depressed when compared with performance on live diets [39, 40, 41]. Zebrafish are gape-limited predators, particularly during the larval stage. First feeding zebrafish have a maximal gape size of <"100 μm and prey items must accordingly be within a similar range. Paramecium (150– 200 μm) and rotifers (<"250 μm) are ideally sized. *Artemia nauplii* are much larger, typically on the order of 500× 100–150 μm (L×W), but also can be taken by first feeding larvae, as it is possible to successfully rear zebrafish on them exclusively [42].

Why Zebrafish?

During the 1960s, the zebrafish (*Danio rerio*) became increasingly prominent in scientific studies. It has a number of unique features that make it an excellent model for studying human genetics and disease. The Wellcome Trust Sanger Institute started the zebrafish genome-sequencing project in 2001. The whole genomic sequence of the Zebrafish was published in 2013. It has a genomic length of 1,505,581,940 base pairs and 26,247 protein-coding genes [43]. Zebrafish have more protein-coding genes than any other previously sequenced vertebrate, and their genome has more species-specific genes than human, mouse, or chicken genomes. The Teleost Specific Genome Duplication (TSD) [44] is thought to be responsible for some of this increased gene number. A direct comparison of zebrafish and human protein-coding genes shows a number of unique characteristics. For example, zebrafish orthologues exist for 71.4 per cent of human genes. In turn, at least one human orthologue may be found in 69 per cent of zebrafish genes. 47 per cent of human genes have a one-to-one link with a zebrafish orthologue among the orthologous genes. The second largest orthology class (the'one-human-to-many-zebrafish' class) has human genes that are related with many zebrafish genes, with an average of 2.28 zebrafish genes for each human gene, which likely represents the TSD. The leukaemia inhibitory factor (LIF), oncostatin M (OSM), and interleukin-6 (IL6) genes, for example, have no readily discernible zebrafish orthologues, despite the presence of the receptors lifra, lifrb, osmr, and il6r in the zebrafish genome. It's possible that zebrafish proteins with functionally similar functions to LIF, OSM, and IL-6 exist, but their sequence divergence is too wide to designate them as orthologues. Similarly, the zebrafish genome lacks a BRCA1 orthologue, but does include an orthologue of the BRCA1-associated BARD1 gene, which produces a functionally comparable protein, and a brca2 gene, which is involved in oocyte development, likely due to its role in DNA damage repair. In Zebrafish, 84 per cent of genes linked to human disease have a Zebrafish homologue.

The Zebrafish is a small, strong fish that is less expensive to keep than mice. Mating in Zebrafish is triggered by a break in the environment (numerous other fish just lay eggs in the absence of light). Scientists may analyses hundreds of embryos each week since zebrafish create hundreds of offspring at weekly intervals. They grow at an incredible rate, developing as much as a human embryo does in a month in just one day. Because zebrafish embryos are usually transparent, researchers may examine the development of internal structures with ease. A low-power microscope can reveal every blood vessel in a live Zebrafish embryo. Zebrafish eggs are fertilized and develop outside of the mother's body (precocial development), making it a suitable model organism for studying early development. The Zebrafish possesses the same major organs and tissues as humans as a vertebrate. Internal organs like as muscle, blood, kidneys, and eyes have numerous similarities to human systems. Zebrafish have a one-of-a-kind ability to renew and repair heart muscle [45].

Malignant growth in Zebrafish is a rapidly growing field that has the potential to make substantial contributions to the cancer community [46] By taking advantage of the Zebrafish's high throughput screening capability, effective gene targeting technologies, and the utilization of high-resolution real-time imaging tools in live fish, digestive disorders and digestive physiology can be easily mimicked and researched [47]. Zebrafish are an excellent model for studying the hematological (blood) system, [48] viral and inflammatory disorders, as well as cardiac and skeletal muscle dysfunction. Autism spectrum disorders (ASD), Down syndrome, Angel man Syndrome, and Foetal Alcohol Syndrome are examples of neurodevelopmental disorders caused by defects that arise in the early developmental period. Alzheimer's disease, Amyotrophic Lateral Sclerosis (ALS), and Multiple Sclerosis are all neurological disorders (MS). Schizophrenia, bipolar disorder, and obsessive-compulsive disorder are examples of psychiatric disorders. Zebrafish is an excellent choice to study those neurological disorders [49]. Zebrafish can be used as a model organism to study nervous system injury, such as a spinal cord lesion or a stroke [50].

Zebrafish is being used as a model organism for eye genetics research. The discovery of a number of zebrafish mutations that impact the eye and retina, as well as the subsequent cloning of mutant genes, has revealed cellular, molecular, and physiological mechanisms that are critical for visual system development. The identification of novel genes or novel gene functions will have important implications for our understanding of human eye diseases, pathogenesis, and treatment as genetic manipulations, sophisticated techniques for phenotypic characterization, behavioral approaches, and screening strategies advance [51].

In comparison to short-lived genetic animal models like C. elegans, Drosophila, or even mice, zebrafish have a comparatively long-life span (3–5 years). This could be seen as a technological disadvantage for ageing research. This element, on the other hand, may offer promise in terms of presenting a model of extremely gradual rather than quick decline in certain physiological functions and systems, as seen in people. This could also help researchers better understand the importance of minor environmental influences in the ageing process, which may be difficult to detect in short-lived animals due to their subtle and cumulative effects. The rate of human senescence may be influenced by such environmental influences, as well as acute and significant toxicities. As a shoaling fish, zebrafish exhibit intricate social interactions. Individual and societal behavioural tendencies, as well as the social environment, may have a substantial impact on zebrafish health, cognitive capabilities, and ageing rates, much as they do in humans. It's difficult to exaggerate the value of studying the impact of social roles or strategies on the ageing process in a productive vertebrate model. By capturing spontaneous or induced actions in hundreds of larval fish or dozens of adults, high-throughput image processing tools now present new prospects for conducting behavioural screens in zebrafish. This analytical approach was used to characterise zebrafish circadian rhythms and sleep, and it may now be used to analyse zebrafish social interactions as well [52].

International and National Status of Zebrafish Research

Zebrafish research articles have been published in 103 countries. The top 25 most productive countries are listed below. There were 18 developed economies among the top 25 productive countries. Five countries are emerging economies (People's Republic of China, Singapore, Brazil, Israel, and India), while two are advanced economies (Taiwan and South Korea). According to the World Bank, developed economies are those with a large stock of physical capital and a population that engages in highly specialized activities. Developing economies include countries with low or middle gross national product per capita, as well as five high-income developing economies: Hong Kong (China), Israel, Kuwait, Singapore, and the United Arab Emirates. 12 With 8196 records, the United States is the most productive country, followed by Germany (1871), England (1393), Japan (1284), and the People's Republic of China (1151). With 170 publications, India is ranked 21st. We also looked at the number of zebrafish research institutions in each nation and discovered that Singapore is the most productive, with an average of 12 publications per institution, followed by Scotland with around 10 papers per institution. Harvard University topped the list of the top 25 productive zebrafish research institutes, with 1106 publications, followed by the University of California with 1015, the University of Washington with 705, the Max Planck Society with 693, and the University of London with 541[53]. Noam Miller and Gerlai are well-known scientists who have conducted substantial research on shoaling and schooling in fish [54].

The usefulness of tracking software shines through when it comes to zebrafish behavioral research and basic investigations. There are numerous free and paid software available for tracking zebrafish movements. Freely available tracking software include tox trac and kinovea, while proprietary software includes ethovision and smart [55]. Only a small amount of software has been developed specifically for monitoring numerous zebrafish. When overlapping and crossover happens, the multiple tracking device id tracker claims to minimize object identification loss. Zebrafish is becoming a popular model organism in India. All around India, research studies from diverse sectors are being reported. The first whole-genome sequence of a wild type strain of Zebrafish was reported by the Council of Scientific and Industrial Research Institute of Genomic and Integrative Biology (CSIR-IGIB) in 2009, and the peer-reviewed research publication was later published. Another important CSIR laboratory that has made substantial contributions to Zebrafish research is the Centre for Cellular and Molecular Biology (CCMB). The Indian Zebrafish research community has a wide range of interests, which can be grouped into four general categories: 1) genomics that is functional 2) Development of the embryo 3) diseases that affect humans 4) the field of biotechnology A few research centres are also working on neurodegenerative illnesses. Using genetically tractable model organisms such as Zebrafish, the Indian Institute of Sciences (IISC) in Bangalore has concentrated on understanding the molecular and cellular underpinnings of myopathies and neurodegenerative illnesses. Because Zebrafish are native to the Indian subcontinent, studies using them as a model organism from India would provide a lot of genuinity. Behavioral research on Zebrafish in India, to my knowledge, is extremely scarce. As a result, studies on the behavior of Zebrafish that are both social and scientifically relevant should be encouraged.

The List of Zebrafish Behaviours

The way zebrafish behave in a fish tank might reveal a lot about their health. Active swimming throughout the tank while searching for, finding, and ingesting food is often viewed as indicating fish that are at ease. Zebrafish stay down, near to tank walls, and close together when disturbed or stressed by an external stressor. When a highly unsettling factor causes them to freeze, their swimming speed might increase or decrease dramatically. Individual variation and context-dependent effects are assessed using standardised behavioural studies and advanced quantitative approaches.

In contrast to short-term twitches or spasms caused by sickness, genetic flaws, toxic chemicals, or ageing, abnormal zebrafish body posture can be interpreted in a variety of ways, including contortion of the body for example, droopy tail and bending. This is frequently linked to neurological symptoms and/or neurodevelopmental problems in larval zebrafish. The natural behaviour of zebrafish to congregate is called aggregation. Conspecifics are sought based on chemical, optical, or other cues. This is essentially an exploratory strategy that leads to shoaling, schooling, and/or sexual aggregating behaviour, among other things. Aggression behaviour in zebrafish is a collection of complex actions directed against conspecifics or other objects, such as approaching, fin raising, undulating body movement, mouth opening activity, body colour change, biting, charging, pursuing, and circling. It could appear in the context of defending territory, safeguarding resources, or asserting dominance. The slowness of swimming or loss of normal motor function that is typically found in older zebrafish or after exposure to certain substances, is often accompanied by a droopy tail, and can be measured by a global reduction in distance travelled and/or swimming velocity, is known as akinesia. Alarm reaction can be characterised as an adaptive escape response that acts as an anti-predatory response in the presence of fear-inducing stimulus. It is characterised by increased movement speed and abrupt directional changes, a response that is known as erratic movement. Freezing with frequent opercular movements, alterations in shoaling and diving are all examples of alarm reactions. Anxiety-like behaviour is a collection of complex behaviours elicited by risky or potentially dangerous stimuli or environments. Geotaxis, thigmotaxis, scototaxis, freezing, opercular motions, body colour change, and irregular movements are examples of restricted exploration (zig-zagging). Approach and nibbling behaviours are demonstrated once visual contact with food is established. The display of presence and movement towards an object by zebrafish is known as approach behaviour. Sexual behaviour is an abrupt swimming movement done by females during courtship, independent of any male courtship behaviours.

The opposite of avoidance is exploratory behaviour, which is the approach to an unfamiliar object (part of the boldness trait). Attraction behaviour includes competitive behaviour (*e.g.*, food seeking). Ataxia is a loss of normal body posture and/or coordination of motions (*e.g.*, laying on a side, swimming on a side, corkscrew swimming) caused by neurotoxicity-induced motor incoordination, akinesia, seizure behaviour, and/or paralysis. Zebrafish attack behaviour includes short bursts of fast swimming directed at an opponent, as well as mouth opening and biting are an example of aggression-related behaviour. Attraction is defined as an increase in the amount of time spent near an object or a chemical stimulation, as well as the movement towards that object or stimulus. Increased movement away from an object or a stimulus is referred to as avoidance.

Camouflage reaction and background adaptation is the change in the pigmentation after being exposed to a darker or lighter background, resulting in a darker or lighter look. Melanophore dispersion or aggregations are the causes for these changes. Backward swimming is an abnormal motor behaviour seen in specific times, such as after exposure to certain hallucinogenic substances, despite the fact that it occurs seldom in normal zebrafish (*e.g.*, lysergic acid diethylamide,

LSD). The intermittent style of swimming known as beat-and-glide is characterised by tail thumping followed by gliding, and it usually develops in larvae at 4 dpf. Bending behaviour is a neurological trait that involves swimming with the body bent laterally and can be seen as a part of seizure activity.

Twitch/spasm behaviour is characterised by short-term bouts of this activity. Biting is a fast movement towards a target, involving mouth opening, closing and physical contact. During 'fights,' zebrafish will frequently bite/nip each other around the gill region or fins. Zebrafish may bite/consume any suitably small objects moving within their field of vision at acceptable speeds and this is different from non-aggressive nibbling. Boldness is a personality attribute characterised by increased approach to novel objects, which is often reflected in higher approach to novel objects. Reduced anxiety-like behaviour, body colour change, and greater exploring activity are common in bolder animals.

Body colour change (coloration response) is a phenomenon that occurs when the body's colour changes. A general change in body pigmentation that results in a darker or lighter appearance can be a sign of anxiety, a natural response to lighting/environmental conditions (camouflage response), and is also a part of social behaviour (*e.g.*, display, fight, or courtship), as well as a result of stress/ sickness, or drug-evoked melanophores dispersion (skin darkening) or aggregation. In adult zebrafish, certain medications (*e.g.*, alcohol, ibogaine) cause significant skin darkening, whereas other conditions (*e.g.*, cold exposure, infections) can cause paling (*e.g.*, sickness behaviour). Coloration response can be measured manually (through visual inspection) or automatically (by luminescence).

Buoyancy dysregulation is when the capacity to control buoyancy is hampered. Surfacing, vertical drifting, cartesian diver behaviour, inclined swimming, and tilting are all symptoms of an inability to maintain a constant elevation without exerting physical effort via swimming and is most commonly caused by problems with the swim bladder or other peripheral systems that frequently manifests as surfacing, vertical drifting, Cartesian diver behaviour, inclined swimming, and tilting. An abnormal phenotype that involves alternating between passive vertical drift and sinking is known as Cartesian diver behaviour. Induced in fish by some neuroactive chemicals, such as LSD. Burst swimming is a fast forward swim with significant bend angles that emerges in larval zebrafish at 2 days after hatching. The C-start (C-bend/turn, Mauthner reflex) is a quick escape or startle response in which the fish's body curves to form a C-shape and then pushes itself away from its prior position at an angle with a swift kick. Mauthner cells govern this behaviour, which is seen in both adult and larval fish. Head stimulation induces C-starts in larval zebrafish, but the tail stimulation elicits both C- and S-starts.

Cannibalism is the eating of dead or alive conspecifics (including egg cannibalism). The movement towards a second fish, with increasing acceleration, while the second fish avoids the first, is known as charging. Chemotaxis or chemo attraction is the zebrafish's movement to and preference for specific chemical stimuli that function as chemo attractants. Appetitive olfactory activity (elicited by L-alanine, food extract, and others), chemically mediated kin recognition, and sexual aggregating behaviour are all examples of chemically mediated attraction behaviours (elicited by sex pheromones). When fish follow increasing chemo attractant concentrations, they move faster, and when they locate the signal source, they move slower. Despite the absence of an upper pharyngeal jaw, zebrafish may chew their food by grinding their lower jaw teeth against a chewing pad on the base of the head, a behaviour known as chewing. Repetitive swimming in a circular direction (typically seen during seizures, neurological abnormalities, and following the effect of specific medications) is known as circling. Circling can be quantified manually or through automated video-tracking methods. Some treatments (*e.g.*, glutamatergic antagonists) induce characteristic "tight" circling that can be measured in body length.

Coasting is similar to drifting in that it involves passive sliding without body or fin motions. Coiling is an embryonic movement that describes a full-body contraction that brings the tip of the tail to the head. It can be spontaneous or elicited by touch. Single or alternating left-right bending of the complete body and it can be seen around 18 hpf in the embryo, then steadily reduces in frequency. Zebrafish would avoid colours that cause an intrinsic aversion (*e.g.*, preferring black (scototaxis), or yellow, green, or red vs. blue), which are most likely connected with colours of natural hazards such as predators (this, however, may depend on context). In adult zebrafish, the colour of fish objects plays a significant role in social interaction (*e.g.*, shoaling formation). Swimming in a spiral at a faster pace and in an uncoordinated direction can be usually seen as a part of the seizure phenotype.

Courtship is a series of programmed actions that take place before to spawning. The male will follow or chase the female in a jerky swimming manner, attempting to tail-nose touch (this happens after the male makes visual contact, since first needs to chemically sense the readiness of female and display sexual aggregating behaviour). If an immediate spawning effort fails, the male will often circle (occasionally circling the female) or zig-zag barely above the ground with his body slightly tilted downwards. During this time, the male will make repeated attempts to mate with the female. The female has the ability to approach, escort, present, and lead. The female may chase the male away if the male's advances are undesired. The following endpoints can be measured manually or automatically using video-tracking systems to characterise zebrafish courtship behaviour.

Creeping is a type of slow swimming in which only the pectoral fins drive the fish forward. A single quick acceleration in one direction (*e.g.*, as part of escape behaviour) with the use of the caudal fin is referred to as a darting. It could be a part of a daring or unpredictable movement or zig-zagging pattern (associated with multiple darts, representing fast acceleration bouts in rapid succession, in which the direction of movement also changes in a seemingly stochastic manner between the darts). Dashing is a set of coordinated (propulsive) darting movements that are typically used as an escape response. A natural preference for depth over shallow water is depth preference (however, that shallow water can trigger breeding behaviour).

Dispersion is a rapid escape-like activity in which numerous fish move away from each other before returning with the group and it is generally triggered by a huge (possibly dangerous) moving object, such as a predator. Display is an aggressive social behaviour that is intended to establish dominance or hierarchy and it is involved in fighting. The frontal display occurs when two fish approach each other from the front in the hopes of nipping or biting each other. The movement to or preference towards the bottom of the tank, known as diving or geotaxis, is commonly in response to a threat. Latency to bottom, time in bottom, frequency of visits to the bottom, distance travelled in bottom, and respective top: bottom ratios can all be used to quantify anxiety or avoidance behaviour. Diving is an active, quick, and directed zebrafish action (with the body travelling towards the bottom head first) and it differs from sinking or cartesian diver behaviour (usually happening in horizontal body posture), or resting behaviour, which are passive, slower, and undirected.

In teleost fishes, the dorsal light reflex (DLR) is the tilting of the body axis toward a light source. When the fish is lighted horizontally, it inclines its dorso ventral axis and turns its dorsal surface toward the light source, with the vestibular righting reflex acting antagonistically to correct the fish's body tilt (*i.e.*, the body inclination increases with the illumination intensity but decreases with the gravity). The motor phenotype, droopy tail is linked to neurological abnormalities, akinesia, and global hypolocomotion. Aging, motor deficits, genetic and pharmacological modulations (*e.g.*, exposure to the monoamine-depleting drug reserpine; severe phenotypes may result in inclined swimming) can all cause inclined swimming. Droopy tail is a long-term trait that differs from tail dip. It can also be a typical resting or sleeping activity. Erratic movement is a type of complex activity characterised by abrupt changes in direction or velocity, as well as the repeated rapid darting. It's most commonly seen in adult zebrafish, and it's linked to many darts. Usually elicited by acute stressors (predator exposure, alarm cue release) or represents a general anxiety or fear state and is commonly seen before or after freezing spells and component of the alarm reaction. In response to a fast change in light, larval zebrafish can also exhibit erratic movements. The reaction to fleeing is in response to a shocking stimulus, a considerable body angular acceleration and displacement occurs. A body 'C-bend' (C-start) is the first stage, followed by a contralateral bend and tail beat (s). The initial acceleration is frequently followed by rapid zig-zagging at the tank's bottom and in certain cases, escape can result in jumping. Fast turning is followed by burst swimming in larval fish. Swimming alongside a male or remaining motionless while being courted are examples of escort behaviour used by females during courting. Exploratory activity is a broad category of behaviours aimed at learning about new places. Locomotor activity and anxiety-related characteristics are connected but not dependent on each other (for example, the exploratory profile of zebrafish can be measured by quantifying the ratio of their activity in different horizontal sections and vertical areas of a tank).

Fast turns are characterised by fast, large-angle twists that require bending of the entire body with high angular velocity. Fear-like conduct is the result of this. Anxiety is a condition or response caused by possible aversive stimuli, whereas fear is a direct response to the appearance or perception of such stimuli, according to traditional clinical views. As a result, anxiety becomes more widespread, whereas dread becomes more cue-oriented. Currently, it's unclear that how the two situations translate into zebrafish behaviour, however some (*e.g.*, alarm reaction) appear to represent a pathological anxiety-like state, while others (*e.g.*, retreat) appear to represent a fear-like state. Fighting is an aggressive confrontation between two individuals that is often used to establish social dominance. It is divided into two phases: the fish first assess each other by showing display, biting/nipping, flicking, and circling behaviours, which continues until the first chase or flee occurs. Following that, the 'winner' (chaser) begins all agonistic actions, while the 'loser' flees, submits, or freezes. Swimming in the shape of an eight is referred to as 'figure eight' swimming. A unique swimming motion found in zebrafish after exposure to certain drugs could be a part of natural courtship behaviour (when the male fish swims around the female with raised fins). The acceleration of movement away from another fish or stimulus is known as fleeing or flight behaviour. Flicking is an agonistic behaviour that occurs when two zebrafish swims towards each other, momentarily brush mouths, and then flick away in opposing directions. It can occur multiple times during agonistic encounters (fight).

Floating is the passive swimming (typically near the water surface), as opposed to surfacing (typically a more active locomotion at the water), drifting (typically in the middle of the water layer), or sleep/rest (typically near or at the bottom) and it can be caused by neurological impairments or buoyancy dysregulation. A common type of zebrafish foraging behaviour is food seeking. It is activated by hunger and suppressed by pathogenic circumstances (*e.g.*, sick behaviour) or certain psychiatric medicines that functions as appetite suppressants. Following behaviour is a nonaggressive movement towards another fish that is frequent during courtship and social engagement. Foraging behaviour is defined as the movement of searching and/or probing in response to sensory stimuli (*e.g.*, food seeking). Freezing refers to the fish's complete lack of movement (save for their gills and eyes) while at the bottom of the tank. Freezing can be quantified by examining the latency, frequency, duration, and location of freezing as a result of high stress or anxiety or as part of submissive conduct. During stress-induced freezing, opercular motions (respiration/gill movements) are normally highly frequent. Freezing behaviour is distinct from immobility, which is frequently caused by toxic or sedative drugs (*e.g.*, high ethanol concentrations), and is accompanied by hypolocomotion and akinesia and it can also result in sinking.

Jaw motions are the stereotypical non-foraging mouth opening activity seen after taking certain medications (*e.g.*, hallucinogenic phencyclidine, convulsant agents, such as domoic acid). The tendency to display a markedly diminished reaction after repeated exposure to a novel stimulus or environment is known as habituation. Habituation is divided into two categories: inter-trial and intra-trial. Increased top exploration, reduced diving, and unaltered irregular motions are normal with time. Calculating the ratios of behavioural activity during the beginning vs. latest trials, or examining the behavioural profile of fish across the trial, can be used to quantify zebrafish habituation.

Head-butting is defined as a single or repeated thrust of the head against a vertical surface (*e.g.*, glass, rock) and it is typically noticed during mirror stimulation, thrashing behaviour, or as a side effect of some psychotropic/hallucinogenic substances (*e.g.*, LSD or ibogaine). In larval zebrafish, head shake motions are a sort of seizure/tremor like activity that is frequently accompanied by convulsions (typical for some convulsant drugs, *e.g.*, domoic acid). Hiding is the fish's attempt to hide (for example, under a stationary object/shelter). The tendency for a fish to construct a crucial "secure" spot where it spends more time and returns after investigating a new environment is known as home base formations. Time spent, number of visits, and distance travelled in home base (vs. non-home base) places is all indicators of natural place preference behaviour. Some pharmaceutical interventions may make it sensitive.

Hyperactivity burst is a single behavioural bout of darting-like erratic movements with rapid turning and high velocity locomotion and can be seen in both adult and larval fish, for example, under high anxiety states or as seizure behaviour. Hyper locomotion is abnormally quick swimming that lasts for a long time and it's usually linked to psychostimulant/convulsant action or anxiety-like behaviour. Hypolocomotion is a condition in which a person swims abnormally slow over an extended period of time. It is often associated with drowsiness, neuromotor impairments, and akinesia. Immobility is a total stop of movement (except for gills and eyes) at the bottom of the tank; it differs from freezing and resting in that it is not necessarily coupled with changed opercular movements (either increased or diminished) (note, however, that immobility and freezing are often used as synonyms in zebrafish literature). Sedative substances (such as high ethanol concentrations) can cause hypolocomotion, akinesia, or paralysis in animals. Swimming on an incline is a form of exercise. An abnormal phenotype (swimming at an angle to the water's surface; tilting) caused by neuroactive or neurotoxic chemicals. It's linked to a droopy tail and buoyancy dysregulation, and it's caused by motor deficiencies or ageing.

J-bend is a type of fine reorientation tuning in which the larva's body curves slightly and has a distinct bend at the tail. 'Jittery' swimming is a type of swimming characterised by repeated short 'jerky' movements and poor smoothness of swimming trajectories and it is typical in seizure behaviour and can be generated by certain medications (*e.g.*, RDX, strychnine). Jumping is a zebrafish behaviour that involves jumping out of the water/tank and it is usually induced by anxiogenic cues as a part of an escape or alarm response (*e.g.*, can be triggered by predator or alarm cue exposure). The preference for related zebrafish over unrelated zebrafish, which is non-existent in larvae but very strong in juvenile 21-dpf zebrafish, is based on chemical and visual cues. Kin recognition is the capacity to detect and seek kin (from unrelated zebrafish) based on chemical and visual cues and includes approach and attraction, and will eventually lead to greater time spent near kin. When the usual body posture is lost due to ataxia, which is commonly seen as a result of sedation and/or neurotoxicity-induced motor incoordination, lying on one's side is noted. Leading is the act of females returning to one spot in the tank at least three times during courting. Lethargy is a behavioural state in adult zebrafish that comprises decreased locomotor activity, diminished escape reaction, unusual body colour, and keeping near to the bottom, with fins (particularly dorsal) often held close to the body (similar to a broader term sickness behaviour). The chronic character of social submissive behaviour, as well as the independence of social environment, distinguishes it from social submissive behaviour.

Looping is a distinct circular swimming activity in larvae around a virtual point outside of the larva's body that begins as early as 5 dpf and is common in mutants with visual feedback abnormalities. Magnetic behaviour refers to zebrafish's behavioural response to magnetic fields (*e.g.*, preferred spatial orientation). Meandering is defined as movement without a clear direction or path; it can be caused by increased anxiety, especially when moving erratically. The mirror stimulation response is a combination of complex actions elicited in fish when they are exposed to a mirror and it is most likely linked to aggression and often comprises approach, head-butting, biting (the mirror), and chasing (own reflection).

The inability to coordinate one's movements is known as motor incoordination. A broad loss of normal body movement coordination (*e.g.*, swimming on one side, swimming in a corkscrew); usually seen as a result of neurotoxicity or other neurological abnormalities. Opening of the mouth or frequent mouth opening (different from chewing or biting behaviour) reported after treatment with various medicines, which might be part of aggression/attack, snapping (*e.g.*, when exposed to food scents), or a specific stereotypic behaviour (*e.g.*, jaw motions). During rest/sleep, the rate of mouth opening is greatly reduced. Neophobia is the fear of unfamiliar items or foods (for example, food neophobia), and it can be measured by the latency and time spent/frequency of touching the novel object. Nibbling is a nonaggressive object biting (usually, in food seeking or as a part of object exploration).

Nocifensive (pain-related) behaviour is characterised by increased swimming, escape, and tail-beating reactions in response to noxious stimuli, which may often be experimentally generated in fish by chemical, thermal, or electrical stimulation (phasic stimuli induce tail beating and escape, while tonic stimuli induce rubbing, and increased opercular movements). The O-bend is an orientation movement in which the body of a larval zebrafish twists to shift its swimming orientation. Dark flashes typically elicit this reaction, which is slower and independent of Mauthner cells than C-bend. The odorant-evoked behavioural activity known as the olfactory response is a complex one. The respiration/gill movements of zebrafish can be visualised using slow-mode video recording; bi-directionally modulated by various psychotropic drugs, opercular beat rate is markedly increased during distress (*e.g.*, during stress-induced freezing, where opercular beat rate can serve as an additional index of anxiety) and reduced during resting/sleep. Stereotyped tracking eye movements caused by moving objects across the visual field are known as optokinetic response/reflex (OKR). There are two parts to this movement: a smooth pursuit movement that follows the item and a quick saccadic movement that resets the eyes once the object has departed the visual field.

The locomotion caused by a repetitive moving stimulus presentation (*e.g.*, spinning drum) is known as optomotor response/reflex (OMR), because zebrafish will generally swim in the same direction as the moving pattern. The sinusoidal feature of zebrafish locomotion/swimming is oscillations of locomotor activity, with alternating high and lower-velocity periods (typically with a frequency of 4– 5 min) when exposed to unfamiliar surroundings. The release of eggs by the female during spawning is known as oviposition. During spawning, the male swims alongside the female, in close proximity but somewhat behind her, with his head roughly level the female's operculum.

The state of being unable to move is known as paralysis. All movements, including the eyes, gills/operculum, and fins, is completely stopped (similar to ataxia, but with more severe/global motor dysfunction, and often with an atypical posture, such as laying on the side, floating upside down, or standing vertically). Selected neuroparalyzing drugs or inherited neurological mutations are usually to blame.

Photokinesis (phototaxis) is the general movement in reaction to light, which includes both positive (light seeking/dark avoidance, scotophobia) and negative (light aversion/dark avoidance, scotophobia) responses (light avoidance, dark preference, scototaxis). Zebrafish are sensitive to visible light (positive phototaxis in larval fish; light avoidance in adult fish) and to ultraviolet (UV) light (negative phototaxis) (UV avoidance). Photomotor response (PMR) is a stereotypical series of motor behaviours in embryonic zebrafish in response to light stimulation, in which zebrafish exhibit motor excitation (lasting 5–7 s) with vigorous shaking, followed by a refractory phase in which basal locomotion is suppressed and animals do not respond to another light pulse.

Piping is the act of gulping air at the water's surface; it might indicate discomfort (*e.g.*, hypoxia or toxicity), but it can also occur during depth-related swim bladder volume adjustment (during diving). The tendency for a fish to develop a preferred area where it spends more time is known as place preference. Drugs, frequent administration of food/food scents, social incentive, or natural behaviours or preferences can all be used to elicit it. Polarization is a behavioural trait of adult zebrafish that reflects the degree to which group members move in the same direction; it is strong in zebrafish schools and low in shoals.

Prey catching is a complex behaviour of larval zebrafish that involves spotting the prey (*e.g.*, paramecium) visually or through chemo sensation, tracking it with a sequence of regular turns, forward slow swim, and/or J-turns, and then capturing and ingesting (feeding behaviour). The first bends have modest amplitude and are most noticeable at far-caudal locations; later bends originate more rostrally, have higher amplitude, and are accompanied by increased tail-beat frequency. Inspecting for predators is necessary. An exploratory/bold risk-taking behaviour in fish that is associated with an increased or decreased tendency to approach a predator, possibly to gain information about the predator's identification, specific location, and/or current motivational state. When fish are shoaling, they will frequently leave the shoal, swim towards the predator, and then return to the group. The act of chasing reflections is known as reflection chasing. Selected psychoactive (*e.g.*, hallucinogenic) medications can also provoke a behaviour that includes pursuing one's own reflection (*e.g.*, in the observation tank or as part of the mirror stimulation reaction). In adult and larval zebrafish, resting behaviour is a sleep-related behaviour. In larval zebrafish, typical rest behaviour involves floating with the head down or staying horizontal towards the tank's bottom (immobility, hypolocomotion). A subordinate fish swimming fast away from an opponent (*e.g.*, a dominant fish) in response to an attack (*e.g.*, following a strike, bite, chase, or charge), part of fleeing behaviour, is a social activity relevant to dominance in zebrafish.

Both larval and adult zebrafish exhibit rheotaxis, a common behaviour in aquatic organisms that includes turning towards a current and a desire to travel upstream. In an experimental situation, this manifests as zebrafish avoiding the sucking source (*e.g.*, sucking pump or standpipe). Routine turning is a slow spontaneous turn with a significant bend angle that causes the larva to reorient before forward swimming and prey capture; it lacks the huge counter-bend and just a tiny section of the tail bends. School Formation is a relatively polarised group (school) in which many fish travel together in a coordinated/synchronous manner; it is a type of aggregation behaviour that improves feeding efficiency and the capacity to identify and/or avoid predators. Schools may disperse into shoals with less polarisation.

Ataxia, corkscrew (spiral) swimming, hyperactivity, circling, spasms, weavering, head shake motions, tremor, and/or frenetic motility are all examples of seizure-like behaviour seen in both larval and adult zebrafish. Death is a possibility in severe situations. It can be measured manually (using a seizure scale) or automatically (using video-tracking software, by assessing the velocity and distance travelled). Shoaling is the formation of a nonpolarized group (shoal) of adult zebrafish held together by social factors (rather than individual attraction to an external stimulus); it is a type of aggregation behaviour. Anxiety/fear leads the shoal to 'tighten,' or swim closer together, potentially forming a school. The shoal becomes looser and less ordered as a result of hunger or habit. The average inter-fish distance, shoal area size, proximity (time each member of the shoal spent within a specified distance from each other), nearest and farthest neighbour distances, time spent in shoal, and time spent away from shoal can all be measured manually or using automated video-tracking systems (reflecting the uniformity of heading).

Reduced exploratory activity, general activity in a novel environment and/or in reaction to stimuli, or risk-taking behaviour is all examples of shyness (opposite to boldness). Sinking is a type of freezing behaviour in which the fish remains still (save for its eyes and gills) but moves its position in the water column (from top to bottom) without moving any of its fins. Hypoactivity, limited exploration, feeding or food seeking, pale body colour, and lethargy are some of the symptoms of illness (or discomfort) (with fins typically held close to the body). Rest behaviour, comprising reversible immobility/hypo locomotion, higher arousal threshold, reduced respiration rate (*e.g.*, opercular motions) and mouth opening frequency, and a compensatory recovery in response to sleep loss, characterises sleep-like behaviour in zebrafish. Brief periods of immobility, frequently with a drooping caudal fin, alternate with active periods of swimming in adult zebrafish; can be rapidly reversed by shocking stimuli like tapping, sound, or a weak electric field. Small bend angles with bend location near the tail describe slow swimming in larval zebrafish.

Snapping is the reflexive opening and closing of the mouth in response to high concentrations of appetitive stimuli. Spontaneous, rapid body movements (typically as a result of neurological/neurotoxic dysfunction, such as seizure) are referred to as twitching. Spatio-temporal stability is the ability to withstand changes in environmental characteristics during exploration of a novel environment, primarily by scaling locomotor activity (*e.g.*, distance travelled) to the size of the environment while maintaining the activity's temporal budgeting; it can be measured using the temporal distribution of locomotion and position in a test tank (*e.g.*, distance travelled, transitions and time spent in each area).

The normal social behaviour of zebrafish is social interaction, which is defined as a reciprocal shift in zebrafish behaviour impacted by the presence or actions of other conspecifics. Fighting/aggression, shoaling/schooling, courting, and spawning are some examples; approach/boldness (social investigation), social recognition, and social preference are further examples. A natural desire to spend time with conspecifics; can be observed as part of shoaling behaviour, kin recognition, social recognition, or a preference for the 'conspecific' compartments of the tank over the 'empty' compartments. It is important to receive social acknowledgement. During breeding, the male zebrafish approaches the female and curls his body around to place his genital pore near to her. The startle reaction is a touch response is an evolutionarily conserved adaptive behaviour in response to sudden, typically aversive stimuli, such as vibration, light, sound, or touching (*e.g.*, touch response). It may include 'C-start' activity, during which coiling and rushing may be observed. Stereotypic conduct is harmful. Other than swimming, zebrafish exhibit a pattern of rigid, repeated behaviours, such as stereotypic mouth opening behaviour following treatment with hallucinogenic substances. Stereotypic swimming is a pattern of inflexible, repeated actions elicited in zebrafish under certain conditions (*e.g.*, swimming from corner to corner). Striking is an aggression-related activity in zebrafish that occurs when the fish swims rapidly toward the opponent without making physical contact. Approach is distinguished by its higher velocity and aggressive (rather than investigative) attitude. Struggling is a larval zebrafish behaviour characterised by longer alternating motor bursts occurring at lower frequency than swimming. When compared to swimming, motor bursts travel in inverse directions along the body during struggling. Following aggressive encounters amongst zebrafish, submissive conduct is a social characteristic. The caudal section of the body is pointed downward, and the submissive fish remains immobile (fins retracted) near the bottom or near the surface of the aquaria.

Surfacing touching is defined as a fish's behaviour at the water's surface that is elicited by a variety of neuroactive medications, primarily serotonergic agonists and glutamatergic antagonists. It may also be linked to buoyancy dysregulation. Swimming is a type of simple zebrafish locomotion that can be classified as either "prolonged," which can last for minutes, or "sustained," which can last for hours. Slow and explosive swimming are both possible in larvae. Swimming on the side, upside down, vertical swimming, and cartesian diver behaviour are examples of abnormal swimming. Swimming on one side refers to the loss of normal body posture caused by ataxia, which is most typically seen as a result of drowsiness and/or neurotoxicity-induced motor incoordination. Swimming upside down is an abnormal phenotype (swimming backwards) that is frequently generated by neuroactive/neurotoxic drugs.

Tail dipping is an agonistic behaviour in which one fish slightly lowers its tail for a short period of time to signal submission and end the confrontation (differs from droopy tail, which is a long-lasting behaviour; Tail beating is characterised by repeated episodes of rhythmic, rostro-caudally progressing peripheral nerve discharges that are alternated between the two sides of the body; The physical manifestation of tail beating can be seen in the side-to-side sweeping of the tail when viewed from above (that can be measured as tail beat amplitude). Tail-nose is the act of touching another fish's side or tail with the nose or head. Males engage in this behaviour during social interactions, particularly courtship. A synchronised leap in reaction to placing on a wet surface (*e.g.*, after jumping out of water); the amplitude (height) and frequency of the leaps can be measured.

The monopolisation and aggressive defence of a designated region in a habitat/tank is known as territorial behaviour (*e.g.*, spawning sites are a commonly defended territory in zebrafish). Trespassers may be chased or bitten by the dominant fish, or they may engage in a 'fight' to challenge its dominance. Thigmotaxis is a zebrafish anxiety indicator that indicates a propensity for remaining close to the edge/side (and avoiding the core open sections). Thrashing is defined as forceful swimming with the caudal fin in contact with the tank's side or bottom wall; swimming against the tank's glass wall that appears as though the fish is attempting to swim through the glass barrier. Top residence is defined as living in the upper part of the tank, which can entail abnormal swimming very close to the water's surface. Touch reaction is a trait of startle behaviour elicited by touch in an embryo, which responds with quick coiling of the trunk bending over the head; it appears about 21 hpf. Tremor behaviour is a shivering-like activity elicited in adult or larval zebrafish by some neurotoxic/convulsant medications, such as domoic acid, which is especially noticeable in the tail area. In both adult and larval zebrafish, turning involves a simple shift in swimming direction. Slow routine turns (without the huge counter bend of escape turns) and quick turns are two types of turning behaviour seen in larval zebrafish (fast, large angle turns characteristic of escape responses).

UV avoidance (negative photokinesis/phototaxis) has been observed in zebrafish larvae as early as 4 days after hatching. Vertical drifting is an abnormal phenotype characterised by passive vertical motion from the bottom to the top, which is frequently generated by neuroactive/neurotoxic chemicals or agents that cause buoyancy dysregulation. Vertical swimming is a phenotype in fish that involves swimming vertically (usually with their heads up at the surface) and is generally generated by neuroactive or neurotoxic drugs, or agents that cause buoyancy dysregulation. In zebrafish, the vestibulo-ocular reflex (VOR) is the compensatory eye movements made in response to linear/angular acceleration and changes in head position in relation to gravity in order to stabilise the retinal image. In zebrafish, these motions are elicited by the vestibular system' semi-circular canals. Withdrawal-related behaviour is a behavioural syndrome seen in zebrafish when they stop using drugs of abuse, and it is characterised by increased anxiety and fear-like behaviour. A tail sweep and circle along a female's torso, mimicking a 'figure eight,' is zig-zagging behaviour in sexual behaviour, which is commonly performed by males during courtship. The irregular motions with several darts known as stress-induced zig zagging occur when the direction of movement changes in a seemingly alternating manner between the darts [56].

Acknowledgement

This work was supported by DST-SERB Early Career Research Grant (ECR/2018/002479) to BR and CSIR-UGC-Junior Research Fellowship (495/2017) to AS.

REFFERENCES

1. Lieschke GJ, Currie PD. Animal models of human disease: zebrafish swim into view. Nature Reviews Genetics. 2007 May;8(5): 353-67.
2. Robinson NB, Krieger K, Khan FM, Huffman W, Chang M, Naik A, Yongle R, Hameed I, Krieger K, Girardi LN, Gaudino M. The current state of animal models in research: A review. International Journal of Surgery. 2019 Dec 1; 72: 9-13.
3. Choi TY, Choi TI, Lee YR, Choe SK, Kim CH. Zebrafish as an animal model for biomedical research. Experimental and Molecular Medicine. 2021 Mar; 53(3): 310-7.

4. Arunachalam M, Raja M, Vijayakumar C, Malaiammal P, Mayden RL. Natural history of zebrafish (*Danio rerio*) in India. Zebrafish. 2013 Mar 1; 10(1): 1-4.

5. Betancur-R R, Wiley EO, Arratia G, Acero A, Bailly N, Miya M, Lecointre G, Orti G. Phylogenetic classification of bony fishes. BMC evolutionary biology. 2017 Dec; 17(1): 1-40.

6. McCluskey BM, Braasch I. Zebrafish Phylogeny and Taxonomy. In The Zebrafish in Biomedical Research 2020 Jan 1 (pp. 15-24). Academic Press

7. Spence R, Fatema MK, Reichard M, Huq KA, Wahab MA, Ahmed ZF, Smith C. The distribution and habitat preferences of the zebrafish in Bangladesh. Journal of fish biology. 2006 Nov; 69(5): 1435-48.

8. Lee SL, Horsfield JA, Black MA, Rutherford K, Gemmell NJ. Identification of sex differences in zebrafish (*Danio rerio*) brains during early sexual differentiation and masculinization using 17α-methyltestoterone. Biology of reproduction. 2018 Aug 1; 99(2): 446-60.

9. Hosseini S, Ha NT, Simianer H, Falker-Gieske C, Brenig B, Franke A, Hörstgen-Schwark G, Tetens J, Herzog S, Sharifi AR. Genetic mechanism underlying sexual plasticity and its association with colour patterning in zebrafish (Danio rerio). BMC genomics. 2019 Dec; 20(1): 1-7.

10. Singh AP, Nüsslein-Volhard C. Zebrafish stripes as a model for vertebrate colour pattern formation. Current Biology. 2015 Jan 19; 25(2): R81-92.

11. Kimmel CB, Ballard WW, Kimmel SR, Ullmann B, Schilling TF. Stages of embryonic development of the zebrafish. Developmental dynamics. 1995 Jul;203(3): 253-310

12. Singleman C, Holtzman NG. Growth and maturation in the zebrafish, Danio rerio: a staging tool for teaching and research. Zebrafish. 2014 Aug 1;11(4): 396-406

13. Parichy DM. The natural history of model organisms: Advancing biology through a deeper understanding of zebrafish ecology and evolution. Elife. 2015 Mar 25; 4: e05635.

14. Lawrence C, Mason T. Zebrafish housing systems: a review of basic operating principles and considerations for design and functionality. ILAR journal. 2012 Jun 1;53(2): 179-91

15. White-James J, McAndrew D, Badman J, McGarry M. Alternative housing for *Xenopus laevis*. Lab animal. 2008 Apr; 37(4): 161-3.

16. Gelfand I, Barak Y, Even Chen Z, Cytryn E, Van Rijn J, Krom MD, Neori A. A novel zero discharge intensive seawater recirculating system for the culture of marine fish. Journal of the World Aquaculture Society. 2003 Sep;34(3): 344-58

17. Watson CA, Hill JE. Design criteria for recirculating, marine ornamental production systems. Aquacultural Engineering. 2006 May 1; 34(3): 157-62.

18. Henry JQ, Lesoway MP, Perry KJ. An automated aquatic rack system for rearing marine invertebrates. BMC biology. 2020 Dec; 18(1): 1-22.
19. Avdesh A, Chen M, Martin-Iverson MT, Mondal A, Ong D, Rainey-Smith S, Taddei K, Lardelli M, Groth DM, Verdile G, Martins RN. Regular care and maintenance of a zebrafish (*Danio rerio*) laboratory: an introduction. JoVE (Journal of Visualized Experiments). 2012 Nov 18(69): e4196.
20. Watanabe T, Kitajima C, Fujita S. Nutritional values of live organisms used in Japan for mass propagation of fish: a review. Aquaculture. 1983 Jul 1; 34(1-2): 115-43.
21. Cahu C, Infante JZ. Substitution of live food by formulated diets in marine fish larvae. Aquaculture. 2001 Aug 15;200(1-2): 161-80
22. Bryant PL, Matty AJ. Optimisation of Artemia feeding rate for carp larvae (*Cyprinus carpio* L.). Aquaculture. 1980 Nov 1; 21(3): 203-12.
23. Bryant PL, Matty AJ. Adaptation of carp (*Cyprinus carpio*) larvae to artificial diets: 1. Optimum feeding rate and adaptation age for a commercial diet. Aquaculture. 1981 Apr 1;23(1-4): 275-86
24. Breder CM, Rosen DE. Modes of reproduction in fishes
25. Eaton RC, Farley RD. Spawning cycle and egg production of zebrafish, *Brachydanio rerio*, in the laboratory. Copeia. 1974 Mar 28: 195-204.
26. Hurk RV, Lambert JG. Ovarian steroid glucuronides function as sex pheromones for male zebrafish, *Brachydanio rerio*. Canadian journal of zoology. 1983 Nov 1;61(11): 2381-7
27. Gerlach G. Pheromonal regulation of reproductive success in female zebrafish: female suppression and male enhancement. Animal Behaviour. 2006 Nov 1;72(5): 1119-24
28. Selman K, Wallace RA, Sarka A, Qi X. Stages of oocyte development in the zebrafish, *Brachydanio rerio*. Journal of morphology. 1993 Nov;218(2): 203-24
29. Darrow KO, Harris WA. Characterization and development of courtship in zebrafish, *Danio rerio*. Zebrafish. 2004 Feb 1; 1(1): 40-5.
30. Spence R, Smith C. Male territoriality mediates density and sex ratio effects on oviposition in the zebrafish, *Danio rerio*. Animal Behaviour. 2005 Jun 1; 69(6): 1317-23.
31. Spence R, Smith C. Mating preference of female zebrafish, *Danio rerio*, in relation to male dominance. Behavioural Ecology. 2006 Sep 1;17(5): 779-83
32. Gerlach G, Lysiak N. Kin recognition and inbreeding avoidance in zebrafish, *Danio rerio*, is based on phenotype matching. Animal Behaviour. 2006 Jun 1; 71(6): 1371-7.

33. Akimenko MA, Johnson SL, Westerfield M, Ekker M. Differential induction of four msx homeobox genes during fin development and regeneration in zebrafish. Development. 1995 Feb 1;121(2): 347-57

34. Laale HW. The biology and use of zebrafish, Brachydanio rerio in fisheries research. A literature review. Journal of Fish Biology. 1977 Feb; 10(2): 121-73.

35. Pack M, Solnica-Krezel L, Malicki J, Neuhauss SC, Schier AF, Stemple DL, Driever W, Fishman MC. Mutations affecting development of zebrafish digestive organs. Development. 1996 Dec 1; 123(1): 321-8.

36. Holmberg A, Schwerte T, Pelster B, Holmgren S. Ontogeny of the gut motility control system in zebrafish Danio rerio embryos and larvae. Journal of Experimental Biology. 2004 Nov 1; 207(23): 4085-94.

37. Jardine D, Litvak MK. Direct yolk sac volume manipulation of zebrafish embryos and the relationship between offspring size and yolk sac volume. Journal of Fish Biology. 2003 Aug; 63(2): 388-97.

38. Dabrowski K, Miller M. Contested paradigm in raising zebrafish (Danio rerio). Zebrafish. 2018 Jun 1; 15(3): 295-309.

39. Goolish EM, Okutake K. Lack of gas bladder inflation by the larvae of zebrafish in the absence of an air water interface. Journal of fish biology. 1999 Nov;55(5): 1054-63

40. Goolish EM, Okutake K, Lesure S. Growth and survivorship of larval zebrafish Danio rerio on processed diets. North American Journal of Aquaculture. 1999 Jul 1;61(3): 189-98

41. Önal UM, Langdon C. Characterization of two microparticle types for delivery of food to altricial fish larvae. Characterization of two microparticle types for delivery of food to altricial fish larvae. 2000;6(3): 159-70

42. Carvalho AP, Araújo L, Santos MM. Rearing zebrafish (Danio rerio) larvae without live food: evaluation of a commercial, a practical and a purified starter diet on larval performance. Aquaculture Research. 2006 Aug; 37(11): 1107-11

43. Howe K, Clark MD, Torroja CF, Torrance J, Berthelot C, Muffato M, Collins JE, Humphray S, McLaren K, Matthews L, McLaren S. The zebrafish reference genome sequence and its relationship to the human genome. Nature. 2013 Apr;496(7446): 498-503

44. Alsop D, Vijayan M. The zebrafish stress axis: molecular fallout from the teleost-specific genome duplication event. General and comparative endocrinology. 2009 Mar 1;161(1): 62-6

45. Poss KD, Wilson LG, Keating MT. Heart regeneration in zebrafish. Science. 2002 Dec 13;298(5601): 2188-90

46. Mione MC, Trede NS. The zebrafish as a model for cancer. Disease models and mechanisms. 2010 Sep 1;3(9-10): 517-23

47. Zhao X, Pack M. Modeling intestinal disorders using zebrafish. Methods in cell biology. 2017 Jan 1;138: 241-70
48. Shafizadeh E, Paw BH. Zebrafish as a model of human hematologic disorders. Current opinion in hematology. 2004 Jul 1;11(4): 255-61
49. Fontana BD, Mezzomo NJ, Kalueff AV, Rosemberg DB. The developing utility of zebrafish models of neurological and neuropsychiatric disorders: A critical review. Experimental Neurology. 2018 Jan 1;299: 157-71
50. Penberthy WT, Shafizadeh E, Lin S. The zebrafish as a model for human disease. Front Biosci. 2002 Jun 1;7(1-3): D1439-53. Kalueff AV, Stewart AM, Gerlai R. Zebrafish as an emerging model for studying complex brain disorders. Trends in pharmacological sciences. 2014 Feb 1; 35(2): 63-75.
51. Fadool JM, Dowling JE. Zebrafish: a model system for the study of eye genetics. Progress in retinal and eye research. 2008 Jan 1; 27(1): 89-110.
52. Yu L, Tucci V, Kishi S, Zhdanova IV. Cognitive aging in zebrafish. PloS one. 2006 Dec 20; 1(1): e14.
53. Kinth P, Mahesh G, Panwar Y. Mapping of zebrafish research: a global outlook. Zebrafish. 2013 Dec 1;10(4): 510-7
54. Miller N, Gerlai R. Quantification of shoaling behaviour in zebrafish (Danio rerio). Behavioural brain research. 2007 Dec 3; 184(2): 157-66.
55. Naiser F, Šmíd M, Matas J. Tracking and re-identification system for multiple laboratory animals. In Visual observation and analysis of Vertebrate And Insect Behaviour Workshop at International Conference on Pattern Recognition, Vol. 0 2018 (pp. 0-0)
56. Kalueff AV, Gebhardt M, Stewart AM, Cachat JM, Brimmer M, Chawla JS, Craddock C, Kyzar EJ, Roth A, Landsman S, Gaikwad S. Towards a comprehensive catalog of zebrafish behaviour 1.0 and beyond. Zebrafish. 2013 Mar 1; 10(1): 70-86.

Chapter 14

Antimicrobial Resistance in Global Aquaculture Settings: A Rising Concern

Reshma Silvester[1], Harikrishnan M.[1] and Mohamed Hatha, A.A[2]*

[1]School of Industrial Fisheries,
[2]Department of Marine Biology, Microbiology and Biochemistry,
School of Marine Sciences,
Cochin University of Science and Technology, Cochin – 682 016, Kerala
**e-mail: sanasilvester@gmail.com*

ABSTRACT

Aquaculture is a rapidly expanding animal production sector and a major contributor to the global demand for animal-based protein food as well. However, with the intensification of the sector, there has been an increased incidence of aquatic diseases globally, causing huge economic loss. Hence, antibiotics are widely used in these farms for disease control and prevention, thus driving the rapid rise of antimicrobial resistance (AMR); a major global health issue. AMR once triggered can easily be transmitted to the surrounding environments and humans/animal pathogens and hence aquaculture farms are widely regarded as potential hotspots for AMR dissemination. Necessary actions have to be taken on an urgent basis to control the unprecedented antibiotic usage in the farms, thereby mitigating the issue of AMR. In addition, the production sectors should focus on shifting to more environmental-friendly alternatives, so as to reduce the dependency on antibiotics.

1. Introduction

In the recent years there has been an increased global demand for fish-based animal protein to meet the food supply of burgeoning population. This is projected to make aquaculture surpass other animal production sectors by 2030 (Schar *et al.,* 2020). However, with rapid intensification of this sector, there is increased incidence of diseases leading to widespread use of antibiotics for prophylaxis and therapeutics as well as for growth promotion. This has in turn resulted in the emergence of antimicrobial resistance (AMR) in aquaculture settings and intermittently brought the industry close to collapse. AMR spreading will be very fast among aquatic microbial communities and it also reaching to human pathogens (Pepi and Focardi, 2021). Thus, commercial fish and shellfish, may act as potential reservoir for multi-drug resistant pathogenic bacteria. AMR is currently a major area of emphasis and work priority of FAO's Department of Fisheries and Aquaculture (FAO, 2017). Hence, AMR surveillance in aquatic animals is crucial as it ultimately harms the human health.

2. Antibiotic Usage in Aquaculture Sector

Antibiotics are widely used in aquaculture production systems for prophylaxis and treatment of bacterial diseases (Romero *et al.,* 2012). They are either mixed in medicated feed or as immersion bath. Antibiotics are poorly absorbed in the fish gut and approximately 70-80 per cent of them are excreted back as undigested food or fish faeces, contaminating the surrounding environmental compartments (Chen *et al.,* 2015). Presence of antibiotic residues in the aquatic environments may change the environmental microbiomes (FAO, 2018). Antibiotics are relatively stable and non-biodegradable, and hence tremains in fish and shellfish meant for consumption (Santos *et al.,* 2016). The presence of antibiotic residues in animal products may also pose severe toxic implications (eg. carcinogenicity effect by chloramphenicol and sulfamethaxine residues) to the consumers (Treiber and Beranek-Knauer, 2021), further aggravating the situation. It ultimately harms the human health; it is crucial to monitor the extent of antibiotic usage (AMU) in this sector.

Developing countries, including India are the major producers of the aquaculture products. However, there is lack of proper monitoring or regulations on AMU in the production systems. Due to this, tonnes of containments of fish and aquaculture products from these countries are been rejected annually in the global markets, due to presence of banned antibiotics such as chloramphenicol and nitrofurantoin (Figure 2). However, in the developed nations; the major consumers of these products, the situation is entirely different. Very few antibiotics are being approved by developed nations and they have stricter regulations for the use of

has its own advantages and limitations, and the choice of method depends on the nature of the farmed fish species, such as size, feeding habits, culture temperature, and economic value (Su *et al.*, 2021).

There are 34 fish vaccines commercially available till date (Su *et al.*, 2021). In 2019, an injectable vaccine against *Flavobacterium* was developed by researchers from University of Stirling, UK to be used in rainbow trout farms (Hoare *et al.*, 2019). In 2021, CIBA developed first indigenous vaccine to be used in fish farms in India. There are few vaccination programs which were successful in aquaculture from various countries. For instance, Norway's fishing industry were able to cut down the usage of antibiotics in their salmon farms to virtually zero, by vaccinating the fishes. Similarly, UK were also able to bring down it's antibiotic usage in their salmon farms through vaccination programs.

4.2 Probiotics

Probiotics are live beneficial microorganisms intended to improve the gut health and provide immunity to the host when consumed. According to WHO/FAO they are defined as "live microorganisms which when administered in adequate amounts confer a health benefit on the host". They are used in the control of many aquatic diseases. Some of the probiotics commonly used against aquaculture pathogens include *Bacillus amyloliquefaciens, B. coagulans, Brevibacillus brevi, Lactobacillus, Bifidobacterium* (*B. bifidum, B. lacti and B. thermophilum*), *Pseudomonas fluorescens* and *Saccharomyces cerevisiae etc.* (Lee *et al.*, 1999). A higher survival and protection against *Aeromonas hydrophila* infection was observed in *Labeo rohita* which was fed with probiotic *Bacillus subtilis* (Kumar *et al.*, 2006). Similarly, better growth was observed in common carp fed with probiotics (Faramarzi *et al.*, 2011).

4.3 Bacteriophage Therapy

Bacteriophage therapy is one of the promising solutions as alternatives to antibiotics for use in aquaculture farms. They are viruses that attack bacteria and do not harm humans or animals. Hence, they are totally safe, reasonably inexpensive and more environment friendly. They are omnipresent in the aquaculture environment, and the total number of viruses normally surpass the bacterial cell concentration by tenfold (Thiel, 2004) and are generally regarded as safe (GRAS). Different modes of phage administration have been used in fish and shellfish; the most common being the parenteral, oral and immersion/bath (Ryan *et al.*, 2011).

Use of bacteriophages isolated from Chilean aquaculture reduced mortality due to *F. psychrophilum* in Atlantic salmon and rainbow trout (Castillo). The world's first bacteriophage biocontrol product, Custus, targeting bacteria causing yersiniosis in salmon was developed in 2019 by a UK based fish health company STIM. Another bacteriophage product to combat *Moritella viscosa*, a cause of winter ulcers in farmed salmon was developed by the same company in 2021. In 2020, Central Institute of Brackish-water Aquaculture (CIBA) developed an indigenous bacteriophage to be used in shrimp farms.

5. Conclusion

There is a widespread emergence of AMR in the aquaculture settings worldwide as evident from the recent reports. Hence, improving awareness among the farmers, strengthening the current surveillance systems, good governance and adoption of good farming practices are some of the urgent strategies required to address AMR in aquaculture, at both national as well as global level.

REFERENCES

Al-Othrubi S. M., Kqueen C. Y., Mirhosseini H., Hadi Y. A., Radu S. 2014. Antibiotic resistance of Vibrio parahaemolyticus Isolated from cockles and shrimp sea food marketed in Selangor, Malaysia. Clin. Microbiol. 3, 148–154.

Bostock J., McAndrew B., Richards R., Jauncey K., Telfer T., Lorenzen K., Little D., Ross L., Handisyde N., Gatward I., *et al.,* 2010. Aquaculture: Global status and trends. Philos. Trans. R. Soc. B. 365, 2897–2912.

Buschmann A.H., Tomova A., Lopez A., Maldonado M.A., Henriquez L.A., Ivanova L., Moy F., Godfrey H.P., Cabello F.C. 2012. Salmon aquaculture and antimicrobial resistance in the marine environment. PLoS ONE. 7, e42724.

Castillo D., Higuera G., Villa M., Middelboe M., Dalsgaard I., Madsen L., Espejo R.T., 2012. Diversity of *Flavobacterium psychrophilum* and the potential use of its phages for protection against bacterial cold-water disease in salmonids. *J. Fish Dis.* 35, 193–20

Chen H., Liu S., Xu X.R., Liu S.S., Zhou G.J., Sun K.F., Zhao J.L., Ying G.G. (2015) Antibiotics in typical marine aquaculture farms surrounding Hailing Island, South China: Occurrence, bioaccumulation and human dietary exposure. Mar. Pollut. Bull. 90, 181–187.

Deepanjali A., Kumar H. S., Karunasagar I., Karunasagar I. 2005. Seasonal variation in abundance of total and pathogenic Vibrio parahaemolyticus bacteria in oysters along the southwest coast of India. Appl. Environ. Microbiol. 71, 3575–3580

FAO (2016a) Fisheries and aquaculture software. FishStatJ—softwarefor fishery statistical time series. In: FAO Fisheries andAquaculture Department, Rome. http: //www.fao.org/fishery/statistics/software/fishstatj/en

FAO. 2017. Towards gender-equitable small-scale fisheries governance and development - A handbook. In support of the implementation of the Voluntary Guidelines for Securing Sustainable Small-Scale Fisheries in the Context of Food Security and Poverty Eradication, by Nilanjana Biswas. Rome. 169 pp. (also available at www.fao.org/3/a-i7419e.pdf)

FAO. 2018. The State of World Fisheries and Aquaculture 2018—Meeting the sustainable development goals. Rome. http: //www.fao.org/3/i9540en/I9540EN.pdf (Accessed 30 July 2019).

Faramarzi M., Kiaalvandi S., Iranshahi F., 2011. The effect of probiotics on growth performance and body composition of common carp (*Cyprinus carpio*). J. Anim. Vet. Adv. 101824082413

Hoare R., Jung S., Ngo T.P.H., Bartie K.L., Thompson K.D., Adams A. 2019. Efficacy of a polyvalent injectable vaccine against Flavobacterium psychrophilum administered to rainbow trout (Oncorhynchus mykiss L.). J. Fish Dis. 42 (2), 229-236

Hinchliffe S., Butcher A., Rahman M.M. 2018. The AMR problem: demanding economies, biological margins, and co-producing alternative strategies. Palgrave Commun. 4, 142.

Kumar R., Mukherjee S.C., Prasad K.P., Pal A.K. 2006. Evaluation of Bacillus subtilis as a probiotic to Indian major crap Labeo rohita(Ham.).Aquaculture Res. 37, 1215–1221

Lee Y.K., Nomoto K., Salminen S., Gorbash S., 1999. Handbook of probiotics New York Wiley.

Letchumanan, V., Yin, W.F., Lee, L.H., Chan, K.G., 2015. Prevalence and antimicrobial susceptibility of *Vibrio parahaemolyticus* isolated from retail shrimps in Malaysia. Front. Microbiol. 6, 33.

Mohammed Kurdi Al-Dulaimi, M., Abd Mutalib, S., Abd Ghani, M., Mohd Zaini, N. A., Ariffin, A. A. 2019. Multiple Antibiotic Resistance (MAR), Plasmid Profiles, and DNA Polymorphisms among *Vibrio vulnificus* Isolates. Antibiotics (Basel, Switzerland), 8(2), 68.

Miranda, C. D., and Rojas, R. 2007. Occurrence of florfenicol resistance in bacteria associated with two Chilean salmon farms with different history of antibacterial usage. Aquaculture 266, 39–46.

Muziasari, W. I., Pitkänen, L. K., Sørum, H., Stedtfeld, R. D., Tiedje, J. M., and Virta, M. 2017. The resistome of farmed fish feces contributes to the enrichment of antibiotic resistance genes in sediments below Baltic Sea fish farms. Front. Microbiol. 7, 2137.

Pepi, M., Focardi, S. 2021. Antibiotic-Resistant Bacteria in Aquaculture and Climate Change: A Challenge for Health in the Mediterranean Area. Int. J. Environ. Res. Public Health 18, 5723

Parthasarathy, S., Das S.C., Kumar, A., Chowdhury, G., Miyoshi, S. I., Dutta, S., Mukhopadhyay, A. K. 2021. Molecular characterization and antibiotic resistance of Vibrio parahaemolyticus from Indian oyster and their probable implication in food chain. World J Microbiol Biotechnol. 37(8), 145.

Romero, J., Feijoó, C. G., and Navarrete, P. 2012. *Antibiotics in Aquaculture-use, Abuse and Alternatives.* Rijeka: INTECH Open Access Publisher.

Ryan, E. M., Gorman, S. P., Donnelly, R. F., and Gilmore, B. F., 2011. Recent advances in bacteriophage therapy: how delivery routes, formulation, concentration and timing influence the success of phage therapy. *J. Pharm. Pharmacol.* 63, 1253–1264

Santos L., Ramos F. 2016. Analytical strategies for the detection and quantification of antibiotic residues in aquaculture fishes: A review. Trends Food Sci. Technol. 52, 16–30.

Schar, D., Zhao, C., Wang, Y. *et al.,* 2021. Twenty-year trends in antimicrobial resistance from aquaculture and fisheries in Asia. Nat. Commun. 12, 5384.

Shah S.Q., Cabello F.C., L'Abée-Lund T.M., Tomova A., Godfrey H.P., Buschmann A.H., Sørum H. 2014. Antimicrobial resistance and antimicrobial resistance genes in marine bacteria from salmon aquaculture and non-aquaculture sites. Environ. Microb. 16, 1310–1320.

Silvester R., Alexander D., Ammanamveetil, M.H.A. 2015. Prevalence, antibiotic resistance, virulence and plasmid profiles of Vibrio parahaemolyticus from a tropical estuary and adjoining traditional prawn farm along the southwest coast of India. Ann Microbiol. 65, 2141–2149

Silvester R., Pires J., Van Boeckel T.P., Madhavan A., Balakrishnan Meenakshikutti A., Hatha M. 2019. Occurrence of dippa-Lactam Resistance Genes and Plasmid-Mediated Resistance Among Vibrios Isolated from Southwest Coast of India. Microb Drug Resist. 25(9): 1306-1315

Su H., Su J. G. 2018. Cyprinid viral diseases and vaccine development. Fish Shellfish Immunol. 83, 84–95

Su H., Yakovlev I. A., van Eerde A., Su J., and Clarke J. L. 2021. Plant-Produced Vaccines: Future Applications in Aquaculture. Front. Plant Sci. 12, 718775.

Tamminen,M., Karkman A., Lõhmus A., Muziasari W. I., Takasu H., Wada S., *et al.,* 2010. Tetracycline resistance genes persist at aquaculture farms in the absence of selection pressure. Environ. Sci. Technol. 45, 386–391.

The World Bank 2017. World Bank Open Data.http: //data.worldbank.org/

Thiel, K. 2004. Old dogma, new tricks—21st century phage therapy. Nature Biotechnology. 1, 31-36.

Thornber K., Verner-Jeffreys D., Hinchliffe S., Rahman M.M., Bass D. and Tyler C.R. 2020. Evaluating antimicrobial resistance in the global shrimp industry. Rev Aquacult, 12: 966-986. https: //doi.org/10.1111/raq.12367

Tomova A., Ivanova L., Buschmann A. H., Rioseco, M. L., Kalsi, R. K., Godfrey, H. P., *et al.,* 2015. Antimicrobial resistance genes in marine bacteria and human uropathogenic *Escherichia coli* from a region of intensive aquaculture. Environ. Microbiol. Rep. 7, 803–809

Treiber F. M., and Beranek-Knauer H. 2021. Antimicrobial Residues in Food from Animal Origin-A Review of the Literature Focusing on Products Collected in Stores and Markets Worldwide. Antibiotics (Basel, Switzerland), 10(5): 534.

Watts J. E. M., Schreier H. J., Lanska L., Hale M. S., 2017. The Rising Tide of Antimicrobial Resistance in Aquaculture: Sources, Sinks and Solutions. Mar Drugs. 15(6), 158.

WHO/FAO 2002. Joint World Health Organization/Food and Agricultural Organization Working Group. Guidelines for the Evaluation of Probiotics in Food 2002. Ontario, Canada

Yang J., Wang C., Shu C., Liu L., Geng J., Hu S., *et al.,* 2013. Marine sediment bacteria harbor antibiotic resistance genes highly similar to those found in human pathogens. Microb. Ecol. 65, 975–981.

Chapter 15

Importance, Production and Use of Egg Yolk Antibody (IgY) in Aquaculture

Thushara James[1], *Mujeeb Rahiman K.M.*[2], *Denoj Sebastian*[3], *Suja Rani S.*[4] *and Dinesh M.D*[5]

[1]*PG and Research Department of Aquaculture and Fishery Microbiology, MES Ponnani College, Ponnani, Malappuram, Kerala*

[2]*School of Industrial Fisheries Cochin University of Science and Technology, Lakeside Campus, Fine Arts Avenue Cochin, Kerala*

[3]*Departmentof Life Sciences, University of Calicut, Calicut University PO, Kerala*

[4]*Departmentof Veterinary Pharmacology and Toxicology, College of Veterinary and Animal Sciences, Mannuthy, Thrissur, Kerala*

[5]*Department of Microbiology, Nehru Arts and Science College, Coimbatore, T.N.*

ABSTRACT

Aquaculture industry has become the only rapidly developing farming sector worldwide; estimated to reach 62 per cent of the total global production by 2030. It is the controlled cultivation of aquatic flora and fauna and providing a protein food source for human consumption. The mass death of farmed fauna or flora owing to illnesses, climate change, and improper handling are among the major issues facing aquaculture production. For a decade, antibiotic treatment was the sole approach to combat predicted microbial illnesses. Improper addition (overdosing), residual difficulties, microbe resistance, and the emergence of new types of microbial species are all adverse effects of antibiotic use in aquaculture. Vaccines, probiotics and prebiotics, among other technologies were developed to address these challenges. The importance of IgY technology, production and various levels of use in aquaculture system is highlighted in this chapter.

Introduction

During 2018, global aquaculture production reached a new high of 114.5 million tons in live weight; with a total farmgate sale value of USD 263.6 billion. The entire aquaculture production consisted of 82.1 million tons of aquatic animals, 32.4 million tons of aquatic algae and 26,000 tons of ornamental seashells and pearls. Inland aquaculture produces 51.3 million tonnes of aquatic animals in the year 2018, accounting for 62.5 per cent of global farmed food fish production (FAO, 2020). Fish and fish products are considered as the World's most traded foods, thus it is not surprise that 67 million tonnes of fish (live weight equivalent) were traded worldwide with an overall export value of USD164 billion. This translates to over 38 per cent of all fish harvested or farmed on the planet. The total amount of fish produced is predicted to increase from 179 million tonnes in 2018 to 204 million tonnes by 2030 (FAO, 2020). Developing countries, mainly in Asia are by far predominant producers with China, Indonesia, India, Viet Nam and Peru being the key producers in 2018.

Aquaculture is a vital part of the economy in both developed and developing countries. It provides food, employment, and income at various levels of business. However, the loss of crops due to climate change, disasters, diseases or other factors which causes financial burden to the persons or company involved directly or indirectly. Disease problems constitute the largest single cause of economic losses in aquaculture sector (Meyer, 1991; El-son *et al.*, 2021). Concurrent with the rapid growth and intensification of aquaculture industry increased use of water bodies, pollution, globalization, and transboundary movement of aquatic fauna, the list of new bacterial pathogens isolated from fish has been steadily increasing (Harvel *et al.*, 1999; Pekala, 2018). Bacterial disease in aquaculture is the major constrain of the cultivation of aquatic species and cause significant economic impact on production globally (Lu *et al.*, 2021; Condon *et al.*, 2022). The disease development is a complicated process, dependent not only on bacterial species being capable of causing health disorders, but also on the immune status of fish, environmental conditions as well as the virulence of the disease agent.

Antibiotic Treatments and its Drawbacks in Aquaculture

Antibiotic usage in aquaculture provides a potentially broader environmental exposure pathway for drug dispersion through water, which has substantial ecosystem health implications as compared to antimicrobial use in terrestrial food animal production (Cabello, 2006; Lulijwa *et al.*, 2020; Holmstrom *et al.*, 2003; Boxall *et al.*, 2004). In addition to disease regulation, the aquatic environmental microbiome's function in nutrient cycling, sustaining biodiversity, carbon sequestration, and freshwater availability remain important research inquiries (Kummerer, 2009). Antimicrobial residues in the aquatic environment change the environmental microbiome and accordingly ecosystem regulatory, provisioning and supporting capabilities (Sarmah *et al.*, 2006; Larsson *et al.*, 2018).

The use of antimicrobials may also serve as reservoirs for antimicrobial resistance genes that provide key routes for human and animal exposure to antimicrobial resistant bacteria (Cabello, 2013; Lulijwa *et al.*, 2020; Larsson, 2018; Heuer, 2009; Aedo *et al.*, 2014; Miller, 2018). Although the mechanisms and directionality of transference can be challenging to establish, shared antimicrobial resistance genes encoded on mobile genetic elements in pathogens isolated from humans, terrestrial and aquatic animals as well as the environment point toward pathways for movement of resistance genes across compartments (Cabello, 2016; Holmstrom *et al.*, 2003; Larsson, 2018; Heuer, 2009; Aedo *et al.*, 2014). Such pathways are illustrated by a putative aquaculture origin for the emergence and dissemination of select plasmid-mediated colistin resistance genes (Shen *et al.*, 2019; Cabello *et al.*, 2017). Increasing use of antimicrobials in humans and food producing animals is driving antimicrobial resistance (Laxminarayan *et al.*, 2013), which is amongst this era's defining global health challenges. Treatment failure rates are increasing as the incidence of antibiotic resistant infections in animal agriculture rises, jeopardising sustainable food animal production and animal welfare.

Egg Yolk Antibodies (EYA) and Mode of Action

Due to the emerging drug resistance in various microorganisms, traditional antibiotic treatment are becoming less effective; thus modern research has mainly focused on developing proficient alternative approaches. Instead of antibiotics for prophylactic and immunotherapeutic use prebiotics, probiotics, various immune stimulating agents and vaccines are the leading alternatives in aquaculture industry. Vaccines get wide acceptance for the fact that there is no risk of drug resistance development in vaccinated animals and protection of minority unvaccinated animals through herd immunity (Gudding, 2014). Out of these valuable therapies, recent studies proved that the application of chicken egg yolk antibodies (EYA) will act as promising alternative mechanism for diagnosis and treatments against various pathogenic organisms and infectious diseases in both humans and animals (Xu *et al.*, 2011; Mohammad *et al.*, 2016; Hu *et al.*, 2019; Paul *et al.*, 2019; Lu *et al.*, 2020; Lee *et al.*, 2021; Shen *et al.*, 2021; Wei *et al.*, 2021). The chicken egg yolk antibodies are very cheaper and easy to produce in large amounts; whereas other polyclonal and monoclonal antibodies are expensive and difficult to produce. This chapter presents an overview on egg yolk antibodies, their salient features, advantages and the potential therapeutic applications of IgY for the prevention and treatment of infectious diseases in aquaculture.

More than 100 years ago, Klemperer (1893) was first demonstrated the avian maternal antibodies are transferred from serum to egg yolk in order to protect the developing embryo from potential pathogens. However, there was no scientific application for this knowledge during that period. When animal welfare became a matter of serious ethical concern for the scientific community, the concepts of Klemperer began to receive more interest. In the year 1996, the workshop of European Center for the Validation of Alternative Methods (ECVAM) also suggested the use of IgY rather than IgG, since then the production and use of IgY technology has become considered an internationally accepted practice (Schade and Hlinak, 1996). Currently, the benefits of IgY technology and its universal application is extensive at large-scale in the field of research, diagnosis, medicine and immunotherapy.

The three classes of immunoglobulins in birds are IgA, IgM and IgY. Immunoglobulin IgY are considered as the human immunoglobulin G (IgG) homolog due to the supposed similarity between the two (Leslie *et al.,* 1969). However, in comparison with IgG, the IgY has a higher molecular weight of 180 kDa than IgG (150 kDa) and also contains a large heavy chain of 65 to 68 kDa (Warr *et al.,* 1995). The major advantage of this higher chain stiffness is that it may be associated with more IgY resistance to proteolyitc degradation as well as fragmentation (Spillner *et al.,* 2012). Furthermore, IgY is more hydrophobic than the IgG and has an iso-electric point between 5.7 and 7.6. In chicken antibodies, often IgG and IgY are interchanged because they are functionally equivalent to each other and also the antibodies in hens serum is IgG and once when it is maternally transferred from hens serum to egg yolk it is called as IgY, which is an egg yolk antibody in high concentrations (Narat, 2003). Antibodies are transferred from hen to chick via the latent stage of the egg and play an important role in immunological function for the relatively immuno-incompetent chick to resist various infectious diseases. Serum IgY is selectively transferred from the hen's circulatory system across the oolemma into the maturing oocyte in the ovarian follicle (Rose *et al.,* 1981).

The term IgY is originally coined to refer to the larger molecular weight IgY found in egg yolk, but it is now accepted that IgY is the major antibody in both the blood and yolk in concentration around 10-20mg/ml (Warr *et al.,* 1995). It makes up about 75 per cent of the total immunoglobulin pool and the serum concentrations of IgY, IgA, and IgM have been reported to be 5.0, 1.25, and 0.61mg/ml, respectively (Leslie *et al.,* 1973). The concentrations of IgA (~0.7 mg/ml) and IgM (~0.15mg/ml) in egg white are relatively low; while that of IgY (8-25 mg/ml) in egg yolk is much high (Rose *et al.,* 1974). A typical egg of an immunized hen contains about 100mg of IgY and a single immunized hen lays approximately 325 eggs per year, resulting in a total production yield of up to 40g of IgY per hen per year (Pauly *et al.,* 2009).

IgY can be purified via a variety of nontoxic methods including water dilution, low pH-induced precipitation, as well as polyethylene glycol, dextran sulfate and xanthan gum-induced precipitations (Akita *et al.,* 1993) or NaCl extraction (Kubickova *et al.,* 2014). Therefore, in terms of environmental impact, IgY can be efficiently collected and processed on an industrial scale without the usage of harsh chemicals and the production of toxic by-products (Yokoyama *et al.*, 1993). Huge quantities of IgY antibodies can be harvested from the eggs laid by a single immunized hen, and the overall yield is comparable to the amount that can be collected from large mammals such as goats and cows and 18-times higher than what is collected from rabbits (Karlsson *et al.*, 2004).

IgY may provide a low cost (only a few cents to produce an oral or nasal dose), safe and fast approach (produced in 5-8 weeks, including the time required for immunogen preparation) for rapid prophylaxis and therapeutic development against a variety of pathogens. In particular, passive immunoprophylaxis by surface treatment with IgY (e.g. via nasal spray, lozenges *etc.*) may be an effective method of capturing pathogens before they enter the body of both humans and animals. This approach may be especially attractive in limited resource countries where vaccination is unavailable, where new viral variants may evade immunity produced by vaccines or previous infection, or where immediate immunity is required (Lee *et al.*, 2021).

IgY used as prophylaxis can also be tailored to be highly specific and is not dependent on the host immune response. This latter feature may enable its clinical application for protecting a wide range of especially vulnerable patients, including the elderly, immune-compromised and young children (Rahman *et al.*, 2013). In addition, in animals raised for conventional antibody production, IgG is the main serum antibody collected. In contrast, IgY antibodies can be easily harvested and used in diagnostic assays, with lower background and cross-reactivity compared with current standards using IgG in animals raised for conventional antibody production.

The exact mechanisms through which IgY counteracts pathogen activity have not been determined. However, several mechanisms have been proposed by Xu *et al.* (2011), Abbas *et al.* (2019) and Paul *et al.* (2019) including agglutination of bacteria (Tsubokura *et al.*, 1997; Thu *et al.*, 2017), inhibition of adhesion (Jin *et al.*, 1998; Lee *et al.*, 2002; Xu *et al.*, 2011), opsonization followed by phagocytosis (Nie *et al.*,2004; Zhen *et al.*, 2008) and toxin neutralization (Wang *et al.*,2011; Thu *et al.*, 2017). Among these, inhibition of adhesion is considered the primary mechanism by which specific IgY functions. Briefly, IgY antibodies generated against intestinal disease causing organisms may reduce the incidence of disease by preventing the attachment of pathogens to the intestine such as blocking the mucosal receptor, interfering with binding to mucins, or neutralizing the colonization factor (Jin *et al.*, 1998).

The IgY Technology Based Production and Purification of IgY Antibodies

Currently, there are several methods of isolation and purification of functionally active chicken IgY from egg yolk described, even for large-scale purification. Nevertheless, purity and yield vary greatly from method to method and require intense optimization for each experiment (Carlander, 2002). The choice of a specific method depends on the yield and purity desired, final application of the IgY as well as material cost, technology, labor skills and scale of extraction. First step is the chicken immunization followed by the yolk IgY extraction, the antibody characterization and finally IgY applications in various kinds of assays (Figure 15.1, Alvaro *et al.*, 2017). Chickens are a suitable model to produce IgY by immunization with bacterial or viral antigens (killed form). To improve the specific IgY response it is demanded the use of adjuvants like Freund's adjuvants for immunization procedures, which is the most frequently used adjuvant. Intramuscularly way is a common way for chicken immunization to produce specific IgY. Alternatively, subcutaneous way provides satisfactory amounts of specific antibody (Behn *et al.*, 2001).

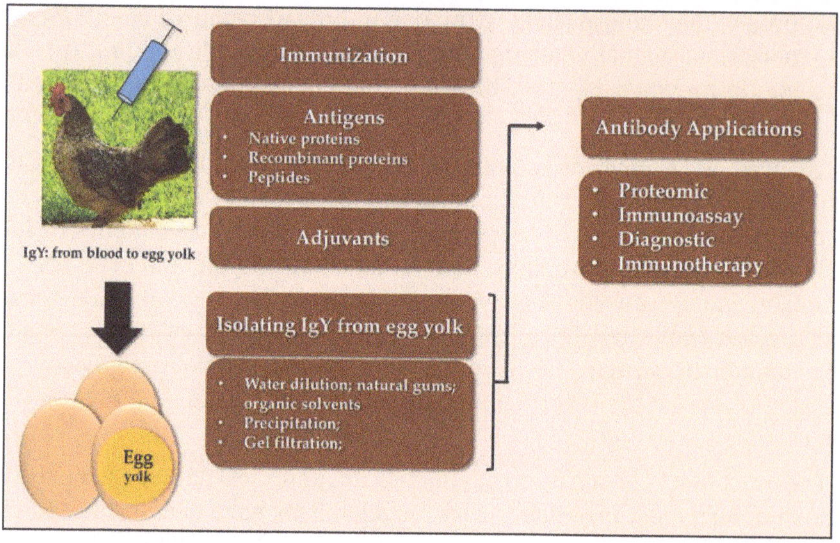

Figure 15.1: General Layout for Producing Specific IgY Antibody.

The immunized Chickens lay daily source of eggs was collected and surfaces sterilized using water or any sterilizing agents. The collected eggs marked properly and stored at suitable temperature for purification. The yolk antibody extraction begins by separating yolk and egg white. After removal of the vitellin membrane, the yolk is put into a measuring cylinder. The next steps in IgY extraction are devoted to removing the lipid content and obtaining a water soluble phase (WSP) antibodies enriched (Lee *et al.*, 2014). The consequent purification procedures vary widely in type and combination and from laboratory to laboratory. The degree application; for example, for oral administration, the whole yolk can be used, and on the other hand, for a reagent, isolated IgY from the mixture of yolk molecules is necessary (Behn *et al.*, 2001).

Removal of lipid content of egg yolk is a very important step in the extraction of IgY. Several natural gums, as carrageenan and xanthan gum, were found to be effective for removal of yolk lipoprotein as a precipitate, resulting in a water soluble fraction with very low lipid concentration (Hatta *et al.,* 1990), Optionally, organic solvents (chloroform) have been used as a mean to remove yolk lipid content at the initial step of IgY isolation. The water dilution method is a widespread procedure for the yolk lipid removal, which provides IgY enriched water soluble fraction (Behn, 2001; Ren *et al.,* 2016).

Several methods can be used for the extraction of IgY from egg yolk and commercial extraction kits are also available (Van Regenmortel, 1993). After the extraction step, three types of separation can be applied to eliminate the contaminants: precipitation with ammonium sulfate $(NH_4)_2SO_4$ (Ko and Ahn, 2007), or sodium sulfate (Wooley and Landon, 1995), or caprylic acid-ammonium sulfate (McLaren *et al.,* 1994), chromatography (Hansen, 1998) and filtration (Hernandez-Campos *et al.,* 2010). Salt based yolk antibody precipitation is promising to be a more efficient and useful purification method for the large scale IgY from egg yolk preparation (Ko and Ahn, 2007). Then IgY precipitated by using $(NH_4)_2SO_4$ and it could be removed by dialysis. Additionally, by following precipitation procedures, the IgY antibodies may be further purified by DEAE cellulose ion-exchange column chromatography (Selvan *et al.,* 2012) or ionic-liquid based aqueous biphasic systems (Taha *et al.,* 2015). The purity could be confirmed by SDS-polyacrylamide gel electrophoresis (Carroll and Stollar, 1983).

Stability and Safety of Chicken Antibodies (IgY)

Even though it is a protein molecule, IgY is resistant to heat and pH, being stable between 30° and 70 °C and active between pH 3.5 and 11. However, the affinity of IgY to its antigen decreases with increasing temperature. One of the most useful characteristics of IgY is its stability during the processing steps and under physiological conditions after administration (Kovacs-Nolan and Mine, 2012). IgY were found to be stable during storage at room temperature or at 4°C for 6 months without losing their activity (Nilsson *et al.,* 2012). Furthermore, IgY antibodies can be stored for 5–10 years at 4°C without significant loss of their activity; they can even retain their activity after storage for 6 months at room temperature or for 1 month at 37°C (Larsson *et al.,* 1997). Even if IgY is a protein molecule, it is resistant to temperature and pH. IgY is being stable between upto 70°C and active between pH 3.5 to 11 (Pereira *et al.,* 2019). According to Shimizu *et al.* (1992) and Hatta *et al.* (1993), IgY is stable at temperature ranging between 60°C and 70°C. IgYs are preferably stored at –20°C for long-term storage; storage at –70°C can cause up to 50 per cent loss of their activity (Staak *et al.,* 2001). The stability of IgY is not affected by freeze-drying, although repeated freeze–thaw cycles might compromise its activity (Shimizu *et al.,* 1988). One report of Fu *et al.* (2006) showed that the thermal stability of IgYs was not affected by lyophilization at 90°C for 15 minutes.

The activity of IgY was decreased at pH 3.5 or lower and almost completely lost with irreversible change at pH 3 (Shimizu *et al.*, 1988). IgY is relatively resistant to trypsin or chymotrypsin digestion, but is fairly sensitive to pepsin digestion. Hatta *et al.* (1993) demonstrated that almost all of the IgY activity was lost following digestion with pepsin, but 39 per cent and 41 per cent of the activity remained even after 8h of incubation with trypsin or chymotrypsin respectively. The stability of IgY against pepsin appears to be highly dependent on pH and the enzyme/substrate ratio. Freezing and freeze-drying are low temperature processes that are usually considered to be less destructive. However, proteins may suffer from a loss of activity as a result of conformational changes, aggregation or adsorption (Skrabanja *et al.*, 1994). However, Shimizu *et al.* (1988) reported freezing and freeze-drying did not affect the activity of IgY unless repeated several times. Fu *et al.* (2006) examined the thermal stability of IgY at various temperatures ranging between 25 and 90°C for 15min treatment, before and after freeze-drying process. Based on their observation, freeze-dried IgY had shown a good thermal stability with no significant reduction in reactive activity except at 90°C, but this was also observed for the non-freeze-dried IgY.

Most of the safety studies are conducted in humans and revealed IgYs are safer than IgGs because they do not bind to human Fc receptors or fix mammalian complement components. Hence they do not trigger potentially dangerous immune responses (Carlander *et al.*, 1999). The researchers found that the levels of pro-inflammatory cytokines tumor necrosis factor (TNF)-dinna, interleukin (IL)-1dippa, IL-6, and granulocyte-macrophage colony-stimulating factor (GM-CSF) were very low in cell cultures treated with IgY compared with the high levels of TNF-dinna and GM-CSF in cells treated with lipo-polysaccharides, indicating that IgYs do not cause inflammatory responses in lung cells and can thus be safely used for prevention of airway infections (Pasteur *et al.*, 2010). Additionally, oral IgY antibodies have been applied to treat rotavirus infections in humans and to treat pulmonary *Pseudomonas aeruginosa* infections, and no negative side effects of IgY treatment have been observed in up to 10 years of use (Nilsson *et al.*, 2007). The safety of intranasal IgY delivery has been documented for the treatment of influenza virus (Nguyen *et al.*, 2010) and acute and chronic pharyngitis for human (Xie *et al.*, 2004).

Advantages of Chicken Antibodies (IgY)

The chicken immune system has been studied for many years, and these studies have contributed substantially to the understanding the fundamental concepts of immunology and the development of different immunoglobulin classes (Carlander *et al.*, 1999). In a report of Sharma (1997), the use of avian antibodies offers many advantages over mammalian antibodies. Nowadays small quantities of avian antibodies are only used because of the limitation of awareness and handling techniques while using chickens. Traditionally, commercially available polyclonal antibodies have been produced in animals such as mice, rats, rabbits, sheep, goats, and horses, and are generally obtained from sera after immunization (Schade *et al.*, 1996). However, these antibodies cannot be prepared on an industrial scale because of the difficulty in obtaining large quantities of blood and concerns about animal welfare (Li *et al.*, 2017). The use of hybridoma technology has been used for the preparation of monoclonal antibodies; however it is still far from the successful commercialization of therapeutic monoclonal antibodies due to the expensive cost.

Recently, the chicken has attracted considerable attention as an alternative source of antibodies. IgY is deposited in the egg yolk in large quantities (Janson *et al.*, 1995), and it can be easily purified from the yolk by simple precipitation techniques, making chickens an ideal source for specific polyclonal antibodies (Gassmann *et al.*, 1990). From the point of view of animal welfare and bioethics, the production of antibodies using a chicken followed by their purification from eggs is more acceptable than preparation of mammalian antisera from blood. Moreover, immunization of the chicken is usually well-tolerated without the formation of abscesses and/or serious health problems, as is common for rabbits. The amount of injected mammalian antigen is often much lower than it is necessary for immunization of rabbits to assure an adequate immune response. Since the antibody is purified from egg yolks, not blood, the stress of the experimental animal is reduced to only the injection of the immunization doses. Animals do not need to be sacrificed to obtain blood for antisera production. Hens are able to produce antibodies with high titers against conserved mammalian proteins.

Hens therefore provide a more hygienic, cost efficient, convenient, and plentiful source of antibodies as compared to the traditional method of obtaining antibodies from mammalian serum (Gassmann *et al.*, 1990; Carlander *et al.*, 2000). Nakai *et al.* (1994) estimated that the productivity of antibodies in hens is nearly 18 times greater than that by rabbits based on the weight of antibody produced per head. Because of the high yolk IgY concentrations, over 100 mg of IgY can be obtained from one egg (Akita and Nakai, 1992). A laying hen produces approximately 20 eggs per month; therefore over 2 g of IgY per month may be obtained from a single chicken (Carlander *et al.*, 1999). In the egg, IgY is stable for months, and once purified it may be stored for years in the cold (Larsson *et al.*, 1993). As the industrial scale automated collection and separation of eggs is currently carried out, the large-scale production of specific IgY for immunotherapeutic purposes is feasible (Cotterill and McBee, 1995).

Applications of IgY Antibodies in Fishes and Aquatic Animals

IgY has been extensively used in health researches, as both therapeutic and diagnostic tool in human and veterinary filed as anti-bacterial, anti-viral, anti-fungal, anti-parasitic, anti-tumour, anti-obesity, anti-allergic and anti-venom activities (Pereira *et al.*, 2019). For the last three decades, IgY based disease diagnosis, prophylactic and disease treatments in aquaculture (Figure 15.2) have been reported and in recent years the development in this field is very promising (Kovacs-Nolan and Mine, 2012; Mohamed *et al.*, 2016; Kumaran *et al.*, 2018; Yi *et al.*, 2018; Eto *et al.*, 2019; Gao *et al.*, 2019; Nakamura *et al.*, 2019; Pereira *et al.*, 2019; Liu *et al.*, 2021; Zhang *et al.*, 2021).

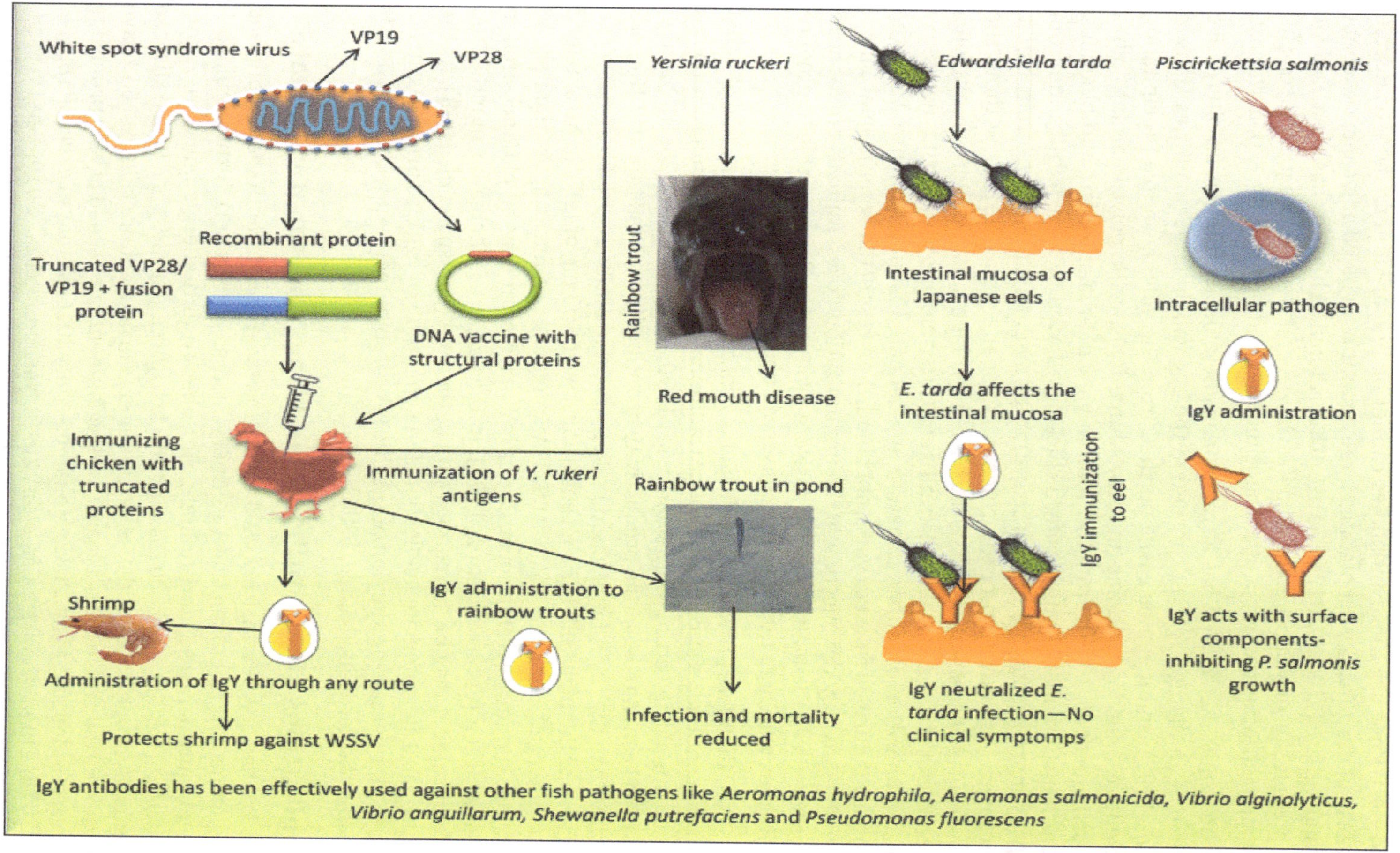

Figure 15.2: Use of IgY in Aquaculture (*Source*: Mohammad *et al.*, 2016).

Application of IgY against Bacterial Diseases

Application of IgY in Crustaceans, Molluscs and Echinoderm

IgY antibodies have been developed against bacterial pathogens such as *Aeromonas hydrophila, A. salmonicida, Vibrio harveyi, V. parahaemolyticus, V. alginolyticus, V. anguillarum, V. splendidus, Enterobacter cloacae etc.*). Passive immunity could possibly protect the crustaceans, molluscs and echinoderms from several diseases including the diseases caused by stress, as they lack an adaptive immune system (Gao *et al.*, 2016a; b; Hu *et al.*, 2019). Study conducted by Li *et al.* (2006) showed anti-*A. hydrophila* IgY protect polyploid gibel carps when challenged with formalin inactivated *A. hydrophila*. The study by Qin *et al.* (2018) showed anti-*A. hydrophila* IgY antibodies significantly increased immunity and maintained the structural integrity after vaccination. Fernandes *et al.* (2019) form their study with anti *A. hydrophila* IgY showed exhibition of specificity and affinity for membrane and cytoplasmic *A. hydrophila* proteins, which help to detect the intra and extra cellular bacterial colonization in the host tissues. Thus, the findings claimed a new perspective on the use of IgY for the diagnosis and analysis of *A. hydrophila* pathogenesis in Nile tilapia (Fernandes *et al.*, 2019). *Aeromonas salmonicida* is a causative agent of ulcers on fish skin. When anti- *A. salmonicida* IgY added in powder form to the water environment of ornamental Koi carp (*Cyprinus carpio* koi), showed good prophylactic activity against the infection and protection from skin ulcer (Gan *et al.*, 2015).

Vibrio spp. are major causes of mortality in shrimp which is lacking adaptive immunity. The study of Gao *et al.* (2016a) and Kumaran *et al.* (2018) showed the protection against *V. harveyi* and *V. parahaemolyticus* on *Litopenaeus vannamei* and *F. indicus*. Toxin-producing *V. parahaemolyticus* caused acute hepatopancreatic necrosis disease (AHPND) to shrimp and anti-PirA-IgY added feed found an effective prophylactic method against this infection. Similar study by Keetanon *et al.* (2021) in the shrimp post larvae 12 (PL 12) of *L. vannamei* (Pacific white shrimp) infected with *V. parahaemolyticus* furnished reduction in mortality rate (25-40 per cent) in contrast to the control group (60 per cent) with IgY-feed proved the effects of IgY on growth, immune responses and pathogenic tolerance.

Experiment conducted with specific and nonspecific anti-*V. alginolyticus* IgY, enhancement of resistance against vibriosis was found in specific IgY feeding experiments and this study also reported the increased survival rate of 90 per cent when fed with specific IgY compared with 10–30 per cent survival rate with nonspecific IgY feed (Wu *et al.*, 2011). In IgY against *V. splendidus* (causing skin ulceration syndrome) was produced and injected intra peritoneal via. 80 per cent of survival rate was occurred in sea cucumber (*Apostichopus japonicus*) and burden of bacteria in IgY treated sea cucumber was also less in the respiratory tract and intestine when compared with untreated control (Li *et al.*, 2016).

Enterobacter cloacae caused mass mortality of *M. rosenbergii* zoeal stages in hatcheries during 2017 in China. Specific egg yolk powder treatment could achieve the survival rate of 67 per cent after E. cloacae infection compared to 20 per cent survival in non-specific egg yolk powder and 17 per cent survival in no egg yolk powder treatment. The damage to hepatopancreas was significantly lower in *M. rosenbergii* treated with specific egg yolk powder than those treated with non-specific egg yolk powder provided a strategy to control the disease by using specific IgY (Gao *et al.,* 2019).

Application of IgY in Fishes

The passive immunization in fish is a good alternative to using antibiotics/chemotherapy, and it is always likely to be successful if the infection occurs through gastro-intestinal tract. Enhancement of phagocytic activity of macrophages for *V. anguillarum* in the presence of specific IgY and decreased levels of tumor necrosis factor-dinna (PaTNF-dinna), interleukin-1dippa (PaIL-1dippa), transforming growth factor-dippa (PaTGF-dippa) and leukocyte cell-derived chemotaxin-2 (PaLECT2) are reported in ayu fih, *Plecoglossus altivelis* against *V. anguillarum* after passive immunization by oral intubation with pathogen-specific IgY (Li *et al.,* 2014). Arasteh *et al.* (2004) reported the protection against *V. anguillarum* by using IgY antibodies through feeding, intraperitoneal injection or oral intubation.

V. harveyi has become a common cause of skin ulceration in Japanese pufferfish in recent years (*Takifugu rubripes*). The first publication on using IgY to manage bacterial illness in pufferfish found that prophylactic and therapeutic effects were achieved through oral administration and immersion of IgY and found improved macrophage phagocytosis of *V. harveyi* (Zhang *et al.,* 2021). The specific IgY incorporated feeding trial in carp (*Cyprinus carpio*) exhibited significant bacteriolytic effects on the non-01 *V. cholerae* pathogen and control the disease (Gao *et al.,* 2017).

Loma salmonae is a microsporidian parasite infecting Pacific salmon and a polyclonal IgY preparation is found to be useful of detecting this parasite in Pacific salmon (Young *et al.,* 2007). Oliver *et al.* (2015) developed IgY against intracellular bacteria *Piscirickettsia salmonis* affecting salmonids causing the disease Piscirickettsiosis. After treating the cell line with specific IgY for 1 h, it was found that the cell line remained uninfected for up to 10 days, when infected with bacteria. Gutierrez *et al.* (1993) reported the neutralizing efficacy of IgY antibodies through oral route against *Edwardsiella tarda* infection and proposed that when Edwardsiellosis/paracolo disease is anticipated, daily administration of anti-*E. tarda* IgY antibodies could be a prophylactic measure. The anti-*E. tarda* IgY antibody test confirmed the prevention of paracolo disease by anti-*E. tarda* IgY antibody (Hatta *et al.,* 2008). Enhancement of passive immunity via feeding microencapsulated specific IgY against Edwardsiellosis, an extremely harmful disease can be caused by *Edwardsiella tarda* suggested the use of this IgY as a valuable preventative in turbot against *E. tarda* 2CDM001 infections (Xu *et al.,* 2020). Experiment conducted by Lee *et al.* (2000) in rainbow trout (*Oncorhynchus mykiss*) fed with anti-*Yersinia ruckeri* IgY antibodies showed lower mortalities after 8 days and found lower number of infection in organs and intestine caused *Y. ruckeri* (pathogen of enteric redmouth disease). Pseudotuberculosis is another fish disease caused by *Photobacterium damselae* subsp. piscicida (Phdp) and Eto *et al.* (2019) from their study reported that anti-*Photobacterium damselae* IgY inhibited the dissemination of *Photobacterium damselae* that could be helpful for the treatment and prophylaxis of this disease.

Application of IgY against Viral Diseases

White spot syndrome virus (WSSV) infection is a major cause of Shrimp mortality in hatcheries and farms. The study of Kim *et al.* (2004) *reported the* survival rate of 50 per cent at day 15 post-challenge with WSSV against shrimp treated with 0.01 mg/10 μl of IgY (produced against the WSSV protein) and suggested these IgY as a potential immunotherapeutant to prevent WSSV infection in *P. chinensis.* By using specific IgY from the hens that had been immunized with inactivated WSSV to neutralize WSSV, the shrimp showed high survival rate (73.3 per cent) contrary to those hens that were immunized with DNA vaccine (33.3 per cent) (Lu *et al.*, 2008). Decreased mortality rate of 20 per cent was observed in crayfish treated with IgY antibodies from inactivated WSSV when compared with 80 per cent mortality with DNA vaccine (Lu *et al.*, 2009) suggested the protection of crayfish (*Procambius clarkiaii*) against the WSSV. The results of Kumaran *et al.* (2010) suggested the use of IgY is an excellent biological source of medicine and prophylactic agent against WSSV infection in shrimps.

Carps are highly infected by Cyprinid herpesvirus 3 (CyHV-3), causes fatal disease with high mortality rate and it was reported that anti-CyHV-3 serum IgY from eggs could prevent the infection of carp brain cell.0Liu *et al.* (2021) from their study suggested the effectiveness of groupers from red-spotted grouper nervous necrosis virus (highly infectious and lethal, mainly infects marine fish larvae) and found 70 per cent protection against this virus infection when fed with specific egg yolk powder.

Conclusion

IgY technology has been extensively used in health researches, as both therapeutic and diagnostic tool in human and veterinary filed. However, the development of IgY technology is in a developing stage and there are scope for various diagnostic and therapeutic IgY antibodies in fisheries and aquaculture sector. This technology can be used in aquaculture or preventing and treating bacterial, viral and parasitic infection diagnosis, tumor diagnosis, hematological tests, various enzyme detection assays, identification of substances *etc.* Besides, IgY represents significant therapeutic resource instead of antibiotic application and also resource for controlling emerging viral infections for which there is no treatment.

REFERENCES

Abbas, A. T., El-Kafrawy, S. A., Sohrab, S. S., and Azhar, E., 2019. IgY antibodies for the immunoprophylaxis and therapy of respiratory infections. Human vaccines and immunotherapeutics, 15(1), 264–275. about: blank

Aedo, S., Ivanova, L., Tomova, A., Cabello, F.C., 2014. Plasmid-related quinolone resistance determinants in epidemic *Vibrio parahaemolyticus*, uropathogenic *Escherichia coli*, and Marine bacteria from an aquaculture area in Chile. Microb. Ecol. 68: 324–328.

Akita, E.M., Nakai, S., 1992. Immunoglobulins from egg yolk: Isolation and purification. J. Food Sci. 57, 629-634.

Akita, E.M., Nakai, S., 1993. Production and purification of Fab' fragments from chicken egg yolk immunoglobulin Y (IgY). Journal of Immunological Methods. 162, 155164.

Alvaro F.J., Jandra, P.S., Paula, B.B., Joely, F.F.B., Eustaquio, R.B., 2017. IgY-Technology Applied to Studies of Toxoplasma gondii Infection. In Toxoplasmosis, IntechOpen, DOI: 10.5772/67997.

Arasteh, N., Aminirissehei, A.H., Yousif, A.N., Albright, L.J., Durance, T.D., 2004. Passive immunization of rainbow trout (*Oncorhynchus mykiss*) with chicken egg yolk immunoglobulins (IgY). Aquaculture. 231, 23–36.

Behn, I., Erhard, M., Hlinak, A., Staak, C., 2001. Chicken Egg Yolk Antibodies, Production and Application – IgYTechnology. Berlin Heidelberg: SpringerVerlag. p. 255.

Boxall, A.B.A., Fogg, L.A., Blackwell, P.A., Kay, P., Pemberton, E.J., Croxford, A., 2004. Veterinary medicines in the environment. In *Reviews of Environmental Contamination and Toxicology*. 180, 1–91.

Cabello, F.C, Tomova, A., Ivanova, L., Godfrey, H.P., 2017. Aquaculture and mcr colistin resistance determinants. MBio. 8, e01229-17.

Cabello, F.C., 2006. Heavy use of prophylactic antibiotics in aquaculture: A growing problem for human and animal health and for the environment. Environ. Microbiol; 8, 1137–1144.

Cabello, F.C., Godfrey, H.P., Buschmann, A.H., Dölz, H.J., 2016. Aquaculture as yet another environmental gateway to the development and globalisation of antimicrobial resistance. Lancet Infect. Dis. 16. PE127–E133.

Cabello, F.C., Godfrey, H.P., Tomova, A., Ivanova, L., Dolz, H.J., Millanavo, A., Buschmann, A.H., 2013. Antimicrobial use in aquaculture re-examined: Its relevance to antimicrobial resistance and to animal and human health. Environ. Microbiol. 15(7), 1917-1942.

Carlander, D., Kollberg, H., Wejaker, P.E., Larsson, A., 2000. Peroral immunotherapy with yolk antibodies for the prevention and treatment of enteric infections. Immunol. Res. 21, 1-6.

Carlander, D., Stalberg, J., Larsson. A., 1999. Chicken antibodies: a clinical chemistry perspective.Ups. J.Med. Sci. 104, 179-189.

Carlander, D., Wilhelmson, M., Larsson, A., 2002. Immunoglobulin Y Levels in Egg Yolk Upsaliensis, Comprehensive Summaries of Uppsala Dissertations from the Faculty of Medicine 1119, pp. 1–15.

Carroll, S.B., Stollar, B.D., 1983. Antibodies to calf thymus RNA polymerase II from egg yolks of immunized hens. The Journal of Biological Chemistry. 258, 2426

Condon, K., Roger Huerlimann., Tony Charles., Ryan Lowrey., Phoebe Arbon., Dean R. Jerry., 2022. Using bacterial whole genome sequencing to identify toxin genes associated with disease outbreaks in black tiger shrimp (*Penaeus monodon*) aquaculture production. Aquaculture. 546, 737255.

Cotterill, O.J., McBee, L.E., 1995. Egg breaking, In Egg Science and Technology, Fourth Edition, eds Stadelman, W. J. and Cotterill, O. J. The Haworth Press, Inc., New York p 231.

El-Son, M.A.M., Nofal, M.I., Hany, M.R., Abdel-Latif., 2021. Co-infection of *Aeromonas hydrophila* and *Vibrio parahaemolyticus* isolated from diseased farmed striped mullet (*Mugil cephalus*) in Manzala, Egypt – A case report, Aquaculture. 530, 735738.

Eto, SF., Fernandes, D.C., Yunis-Aguinaga, J., Claudiano, G.S., Shimada, M. T., Rogério Salvador, de Moraes, F.R., de Moraes, J.R.E., 2019. Characterization and production of IgY antibodies anti-Photobacterium damselae subsp. piscicida: Therapeutic and prophylactic use in Rachycentron canadum, Aquaculture, 513, 734424.

FAO., 2020. The State of World Fisheries and Aquaculture. Sustainability in action. Rome.

Fernandes, D.C., Eto, S.F., Michelli I.G., Funnicelli, Camila C., Fernandes, Ives Charlie-Silva., Belo, M.A.A., João M., Pizauro, 2019. Immunoglobulin Y in the diagnosis of *Aeromonas hydrophila* infection in Nile tilapia (*Oreochromis niloticus*), Aquaculture. 500, 576-585.

Fu, C.Y., 2006. Preparation and evaluation of anti-SARS coronavirus IgY from yolks of immunized SPF chickens. J. Virol. Methods. 133, 112-115.

Gan, H., He, H., Sato, A., Hatta, H., Nakao, M., Somamoto, T., 2015. Ulcer disease prophylaxis in koi carp by bath immersion with chicken egg yolk containing anti-*Aeromonas salmonicida* IgY, Res. Vet. Sci. 99,82–86.

Gao, X., Nan Chen., Yue Zhang., Xiaojun Zhang., Xuwen Bing., 2017. Non-O1 *Vibrio cholerae* pathogen from *Cyprinus carpio* and control with anti-non-O1 *V. cholerae* egg yolk powder (IgY), Aquaculture. 479, 69-74.

Gao, X., Zhang, H., N.,Jiang, Q., Chen, Li, X., Liu, X., Yang, H., Wei, W., Zhang, X., 2019. *Enterobacter cloacae* associated with mass mortality in zoea of giant freshwater prawns *Macrobrachium rosenbergii* and control with specific chicken egg yolk immunoglobulins (IgY), Aquaculture, 501, 331-337.

Gao, X., Zhang, X., Lin, L., Yao, D., Sun, J., Du, X., Li, X., Zhang, Y., 2016a. Passive Immune-Protection of *Litopenaeus vannamei* against *Vibrio harveyi* and *Vibrio parahaemolyticus* Infections with Anti-Vibrio Egg Yolk (IgY)-Encapsulated Feed. International Journal of Molecular Sciences. 17(5), 723.

Gao, X., Zhang, X., Sun, J., Du, X., Li, X., Zhang, Y., Lin, L., 2016b. Passive protection effect of anti-Vibrio anguillarum IgY-encapsulated feed on half-smooth tongue sole (*Cynoglossus semilaevi*) against *V. anguillarum*. Fish Shellfish Immun. 56,483–488.

Gassmann, M., Thomes, P., Weiser, T., Hubscher, U., 1990. Efficient production of chicken egg yolk antibodies against a conserved mammalian protein. Faseb J. 4: 2528.

Gudding, R., 2014. Vaccination as a Preventive Measure. In: Gudding R., Lillehaug A., Evensen O., editors. *Fish Vaccination.* pp. 12–21.

Gutierrez, M.A., Miyazaki, T., Hatta, H., Kim, M., 1993. Protective properties of egg yolk IgY containing anti-*Edwardsiella tarda* antibody against paracolo disease in the Japanese eel, *Anguilla japonica* Temminck and Schlegel. J. Fish Dis. 16, 113-122.

Hansen, P., Scoble, J.A., Hanson, B., Hoogenraad, N.J., 1998. Isolation and purification of immunoglobulins from chicken eggs using thiophilic interaction chromatography. Journal of Immunological Methods. 215, 1–7.

Harvell, C.D., Kim, K., Burkholder J.M., 1999. Emerging marine diseases—climate links and anthropogenic factors. Science. 285(5433), 1505–1510.

Hatta, H., Kapoor, M.P., Juneja, L.R., 2008. Bioactive components in egg yolk. In: Mine Y (ed) Egg Bioscience and Biotechnology, John Wiley Sons Inc., Hoboken, NJ. 5,185–237.

Hatta, H., Tsuda, K., Akachi, S., Kim, M., Yamamoto, T., 1993. Productivity and some properties of egg yolk antibody_IgY.Against human rotavirus compared with rabbit IgG.Biosci.Biotechnol.Biochem.57: 450.

Hatta, H., Kim, M., Yamamoto, T., 1990. A novel isolation method for hen egg yolk antibody, "IgY". *Agric. Biol. Chem.* **54**, 2531-2535.

Hernández-Campos, S.J. Brito-De la Fuente E, Torrestiana-Sánchez B., 2010. Purification of egg yolk immunoglobulin (IgY) by ultrafiltration: effect of ph, ionic strength, and membrane properties. Journal of Agriculture and Food Chemistry. 58, 187-193.

Heuer, O.E., Kruse, H., Grave, K., Collignon, P., Karunasagar, I., Angulo, F.J., 2009. Human health consequences of use of antimicrobial agents in aquaculture. Clin. Infect. Dis. 49, 1248–1253.

Holmström, K., Gräslund, S., Wahlström, A., Poungshompoo, S., Bengtsson, B.-E., Kautsky, N., 2003. Antibiotic use in shrimp farming and implications for environmental impacts and human health. Int. J. Food Sci. Technol. 38, 255–266.

Hu, B., Yang, X., Guo, E., Zhou, P., Xu, D., Qi, Z., Deng, L., 2019. The preparation and antibacterial effect of egg yolk immunoglobulin (IgY) against the outer membrane proteins of *Vibrio parahaemolyticus.* J Sci Food Agric. 99, 2565–2571.

Janson, A.K., Smith, C.I., Hammarstrom, L., 1995. Biological properties of the eggs yolk immunoglobulins. Adv. Exp. Med. Biol. 371, 685-690.

Jin, L.Z., Samuel, K., Baidoo, K., Marquardt, R.R., Frohlich, A.A., 1998. *In vitro* inhibition of adhesion of enterotoxigenic *Escherichia coli* K88 to piglet intestinal mucus by egg yolk antibodies. FEMS Immunol Med Microbiol. 21, 313–321.

Karlsson, M., Kollberg, H., Larsson, A., 2004. Chicken Igy: Utilizing the Evolutionary Advantage. Worlds Poult Sci J. 60, 341–8.

Keetanon, A., Niti Chuchird, Hae-Duck Bae, Mi-Kyoung Won, Su-Yeun Kim, Fazle Elahi., 2021. Effects of IgY Antibody on Growth, Survival, Immune Responses and Protection against *Vibrio parahaemolyticus* in Pacific White Shrimp. Journal of Fisheries and Environment. 45(1), 1-7.

Kim, D.K., Jang, I.K., Seo, H.C., Shin, S.O., Yang, S.Y., Kim, J.W., 2004. Shrimp protected from WSSV disease by treatment with egg yolk antibodies (IgY) against a truncated fusion protein derived from WSSV, Aquaculture. 237(1–4), 21-30.

Klemperer, F., 1893. Ueber natürliche Immunität und ihre Verwerthung für die Immunisirungstherapie. Archiv für Experimentelle pathologie und Pharmakologie. 31, 356–382.

Ko, K.Y., Ahn, D.U., 2007. Preparation of immunoglobulin Y from egg yolk using ammonium sulfate precipitation and ionexchange chromatography. Poultry Science. 86, 40040.

Kovacs-Nolan, J., Mine, Y., 2012. Egg yolk antibodies for passive immunity. Annu. Rev. Food Sci. Technol. 3, 163–182.

Kubickova, B., Majerova, B., Hadrabova, J., Noskova, L., Stiborova, M., Hodek, P., 2014. Effect of Chicken Antibodies on Inflammation in Human Lung Epithelial Cell Lines. Neuro Endocrinol Lett. 35, 99–104.

Kumaran, T., Thangaswamy, S., Citarasu, T., 2010. Production of Anti WSSV IgY Edible Antibody Using Herbal Immunoadjuvant *Asparagus racemosus* and its Immunological Influence against WSSV Infection in Penaeus monodon. J. Aquac. Feed Sci Nutri. 2 (1), 1-15.

Kumaran, T., Thirumalaikumar, E., Lelin, C., Palanikumar, P., Michaelbabu, M., Citarasu, T., 2018. Physicochemical properties of anti Vibrio harveyi egg yolk antibody (IgY) and its immunological influence in Indian white shrimp Fenneropenaeus indicus. Fish Shellfish Immunol. 74, 349-362.

Kümmerer, K., 2009. Antibiotics in the aquatic environment—A review—Part I. Chemosphere. 75, 417–434.

Larsson, A., Balow, R.M., Lindahl, T.L., Forsberg, P.O., 1997. Chicken antibodies: taking advantage of evolution. A review.Poultry Science. 72, a 1807-1812.

Larsson, A., Balöw, R-M., Lindahl, T.L., Forsberg, P-O., 1993. Chicken antibodies: taking advantage of evolution. A review. Poultry Science. 72, 1807–1812.

Larsson, D.G.J., Andremont, A., Bengtsson-Palme, J., Brandt, K.K., de Roda Husman, A.M., Fagerstedt, P., Fick, J., Flach, C.F., Gaze, W.H., Kuroda, M., Kvint, K., Laxminarayan, R., Manaia, C.M., Nielsen, K.M., Plant, L., Ploy, M.C., Segovia, C., Simonet, P., Smalla, K., Snape, J., Topp, E., van Hengel, A.J., Verner-Jeffreys, D.W., Virta, M.P.J., Wellington, E.M., Wernersson, A.S., 2018. Critical knowledge gaps and research needs related to the environmental dimensions of antibiotic resistance. *Environ. Int.* 117, 132–138.

Laxminarayan, R., Duse A., Wattal, C., Zaidi, A.K., Wertheim, H.F., Sumpradit, N., Vlieghe, E., Hara, G.L., Gould, I.M., Goossens, H., Greko, C., So, A.D., Bigdeli, M., Tomson, G., Woodhouse, W., Ombaka, E., Peralta, A.Q., Qamar, F.N., Mir, F., Kariuki, S., Bhutta, Z.A., Coates, A., Bergstrom, R., Wright, G.D., Brown E.D., Cars, O., 2013. Antibiotic resistance—The need for global solutions. Lancet Infect. Dis. 13, 1057–1098.

Lee, HY, Abeyrathne, E.D., Choi, I., Suh, J.W., Ahn, D.U., 2014. Seequential of immunoglobulin Y and phosvitin from chicken egg yolk without using organic solvents. Poultry Science. 93, 2668-2677.

Lee, K.A., Chang, S.K., Lee, Y.J., Lee, J.H., Koo, N.S., 2002. Acid stability of anti-*Helicobacter pyroli* IgY in aqueous polyol solution. J. Biochem. Mol. Biol. 35(5), 488–493.

Lee, L., Samardzic, K., Wallach, M., Frumkin, L.R., Mochly-Rosen D., 2021. Immunoglobulin Y (IgY) for Potential Diagnostic and Therapeutic Applications in Infectious Diseases. Frontiers in Immunology. 12, 696003.

Lee, S.B., Mine, Y., Stevenson, R.M.W., 2000. Effects of hen egg yolk immunoglobulin in passive protection of rainbow trout against *Yersinia ruckeri*. Journal of Agricultural and Food Chemistry. 48, 110–115.

Leslie, G.A., Clem, L.W., 1969. Phylogeny of Immunoglobulin Structure and Function: Iii. Immunoglobulins of the Chicken. J Exp Med. 130, 1337–52.

Leslie, G.A., Martin, L.N., 1973. Studies on the secretory immunologic system of fowl III. Serum and secretory IgA of the chicken. J Immunol. 110: 1–9.

Li X, Jing, K., Wang, X., Li, Y., Zhang, M., Li, Z., Xu, L., Wang, L., Xu, Y., 2016. Protective effects of chicken egg yolk antibody (IgY) against experimental *Vibrio splendidus* infection in the sea cucumber (*Apostichopus japonicus*). Fish Shellfish Immunol. 48, 105-11.

Li, C.H., Lu, X.J., Li, D.F., Chen. J., 2014. Passive protective effect of chicken egg yolk immunoglobulins against experimental *Vibrio anguillarum* infection in ayu (*Plecoglossus altivelis*). Fish Shellfish Immunol. 37, 108-114.

Li, T., Zhou, B., Yu, T., Li, N,. Guo, X., Zhang, T., Zhao, J., Xu, L., Li, S., Ma, L., Li, T., Ding, L., Sun, M., Li, D., Yin, J., 2017. Neutralizing scFv Antibodies against Infectious Bursal Disease Virus Isolated From a Nlpa-Based Bacterial Display Library. Ann Adv Chem. 1, 001-011.

Li, X.L., Shuai J.B., Fang, W.H., 2006. Protection of *Carassius auratus* Gibelio against infection by *Aeromonas hydrophila* using specific immunoglobulins from hen egg yolk. J. Zhejiang Univ. Sci. B. 7, 922-928.

Liu, J., Qin. Y., Yan. L., Liu. W., Shi. H., Lu. Y., Liu., X. 2021. Protective effects of egg yolk immunoglobulins (IgY) on juvenile groupers (*Epinephelus fuscoguttatus* × *Epinephelus lanceolatus*) with red-spotted grouper nervous necrosis virus infection, Aquaculture. Volume 545.

Liu, Z., Ke, H., Ma, Y., Hao, L., Feng, G., Ma, J., Liang, Z., Li, Y., 2014. Oral Passive Immunization of Carp *Cyprinus carpio* with Anti-CyHV-3 Chicken Egg Yolk Immunoglobulin (IgY), Fish Pathology. 49(3), 113-120.

Lu, Y., Liu, J., Jin, L., Li, X., Zhen Y., Xue, H., Lin, Q., Xu, Y., 2009. Passive immunization of crayfish (*Procambius clarkiaii*) with chicken egg yolk immunoglobulin (IgY) against White Spot Syndrome Virus (WSSV). Applied Biochem. Biotechnol. 159(3), 750-758.

Lu, Y., Liu, J.J., Jin, L.J., Li, X.Y., Zhen, Y.H., Xue, H.Y., 2008. Passive protection of shrimp against white spot syndrome virus (WSSV) using specific antibody from egg yolk of chickens immunized with inactivated virus or a WSSV-DNA vaccine. Fish and Shellfish Immunology. 25, 604–610.

Lu, Y., Wang, Y., Zhang, Z., Huang, J., Yao, M., Huang, G., Yao, M., Huang, G., Yuanyuan Ge., Zhang, P., Huang, H., Wang, Y., Li, H., Wang, W., 2020. Generation of Chicken IgY Against Sars-Cov-2 Spike Protein and Epitope Mapping. J Immunol Res. 2020: 9465398.

Lu, Z., Ren Gao., Yafei Duan., Rui Han., Wenjie Guo., Xueming Dan., Yanwei Li., 2021. Isolation and genetic characterization of *Flavobacterium columnare* from grass carp, *Ctenopharyngodon idellus*, in China, Aquaculture. 541, 736762.

Lulijwa, R., Rupia, E.J., Alfaro, A.C., 2020. Antibiotic use in aquaculture, policies and regulation, health and environmental risks: A review of the top 15 major producers. Rev. Aquacult. 12, 640–663.

McLaren, R.D., Prosser, C. G., Grieve, R.C.J., Borissenko, M., 1994. The use of caprylic acid for the extraction of the immunoglobulin fraction from egg yolk of chickens immunised with ovine dinna-lactalbumin. In Journal of Immunological Methods. 177, 1-2, 175–184.

Meyer, F.P., 1991. Aquaculture disease and health management. Journal of Animal Science. 69(10), 4201–4208.

Miller, R.A., Harbottle, H., 2018. Antimicrobial drug resistance in fish pathogens. Microbiol. Spectrum. 6(1).

Mohammad, J.Z., Tiwari, R., Sachan, S., Karthik, Malik, Y.S., Dadar, M., Sarwar, M., Sayab, M., Dhama, K., 2016. Avian Egg Yolk Antibodies (IgY) and their Potential Therapeutic Applications for Countering Infectious Diseases of Fish and Aquatic Animals. International Journal of Pharmacology. 12, 760-768.

Nakai S, Li-Chan E, Lo KV (1994). Separation of immunoglobulin from egg yolk, In Egg Uses and Processing Technologies. New Developments, eds Sim, J.S. and Nakai, S. CAB International, Wallingford, UK. P 94.

Nakamura, R., Pedrosa-Gerasmio, I.R., Alenton, R.R.R., Nozaki, R., Kondo, H., Hirono, I., 2019. Anti-PirA-like toxin immunoglobulin (IgY) in feeds passively immunizes shrimp against acute hepatopancreatic necrosis disease. J Fish Dis. 42(8),1125-1132.

Narat, M., 2003. Production of antibodies in chickens. Food Technol Biotechnol. 41 (3), 259-67.

Nguyen, H.H., Tumpey, T.M., Park, H.J., Byun, Y.H., Tran, L.D., Nguyen, V.D., Kilgore, P.E., Czerkinsky, C., Katz, J.M., Seong, B.L., 2010. Prophylactic and therapeutic efficacy of avian antibodies against influenza virus H5N1 and H1N1 in mice. PLoS One. 5.

Nie, R., Wu, D., Hu, G., Zhang, J., Yang, H., Wen, Z., 2004. Effect of specific egg yolk immunoglobulins on phagocytosis by neutrophils. Chin. J. Vet. Med, 12, 23–25.

Nilsson, E., J.Stalberg. J., Larsson. A., 2012. IgY stability in eggs stored at room temperature or at +4°C. Br Poult Sci. 53(1), 42-6.

Nilsson, E., Kollberg, H., Johannesson, M., Wejaker, P.E., Carlander, D., Larsson, A., 2007. More than 10 years' continuous oral treatment with specific immunoglobulin Y for the prevention of *Pseudomonas aeruginosa* infections: a case report. J Med Food, 10, 375–378.doi: 10.1089/jmf.2006.214.

Olga H., 2017. Aquatic AMR Workshop 1: 10-11 April 2017, Mangalore, India, Major Bacterial Diseases Affecting Aquaculture, FAO, UN)

Oliver, C., Valenzuela, K., Silva, H., Haro, R.E., Corte´s, M., Sandoval, R., Pontigo, J.P., C. A´lvarez, J. E. Figueroa, R. Avendan˜o-Herrera., 2015. Effectiveness of egg yolk immunoglobulin against the intracellular salmonid pathogen Piscirickettsia salmonis. J. Appl. Microbiol. 119: 365–376.

Pasteur, M.C, Bilton, D., At, H., 2010. British thoracic Soc B. British thoracic society guideline for non-CF bronchiectasis. Thorax.

Paul, T., Anirban Pal., Sampa Pal., Nilav Aich., 2019. New Dimension in Fish Immunotherapeutics: Avian Egg Yolk Antibody (IgY). Int.J.Curr.Microbiol.App. Sci. 8(7), 86-94.

Pauly, D., Dorner, M., Zhang, X., Hlinak, A., Dorner, B., Schade, R., 2009. Monitoring of Laying Capacity, Immunoglobulin Y Concentration, and Antibody Titer Development in Chickens Immunized With Ricin and Botulinum Toxins Over a Two-Year Period. Poult Sci. 88, 281–90.

Pêkala-Safiñska A., 2018. Contemporary Threats of Bacterial Infections in Freshwater Fish. Journal of veterinary research, 62(3), 261–267.

Pereira, E.P.V., Tilburg, M.F.V., Florean, E.O.P.T., Guedes, M.I.F., 2019. Egg yolk antibodies (IgY) and their applications in human and veterinary health: a review. Int. Immunopharmacol. 73, 293–303.

Qin, Z., Babu, V.S., Li, N., Fu, T., Li, J., Yi, L., Zhao, L., Li, J., Zhou, Y., Lin, L., 2018. Protective effects of chicken egg yolk immunoglobulins (IgY) against experimental Aeromonas hydrophila infection in blunt snout bream (Megalobrama amblycephala). Fish Shellfish Immunol. 78, 26-34.

Rahman, S., Van Nguyen S., Icatlo FC Jr., Umeda, K., Kodama, Y., 2013. Oral Passive Igy-Based Immunotherapeutics: A Novel Solution for Prevention and Treatment of Alimentary Tract Diseases. Hum Vaccin Immunother. 9, 1039–48.

Ren H, Yang W, Thirumalai D, Zhang x, Schade R. (2016). A comparative evaluation of six principal IgY antibody extraction methods. Alternatives to Laboratory Animals: ATLA, 44, 11-20.

Rose MEM., and E. Orlans., 1981. Immunoglobulins in the egg, embryo and young chick. Dev. Comp. Immunol. 5, 15–20.

Rose, M. E., Orlans, E., Buttress, M., 1974. Immunoglobulin classes on the hen's egg: their segregation in yolk and white. Eur. J. Immunol. 4,521–523.

Sarmah, A.K., Meyer, M.T., Boxall, A.B.A., 2006. A global perspective on the use, sales, exposure pathways, occurrence, fate and effects of veterinary antibiotics (VAs) in the environment. Chemosphere. 65,725–759.

Schade, R., Hlinak, A., 1996. Egg yolk antibodies, state of the art and future prospects. ALTEX, 13 (13), 5-9.

Selvan, K., Sentila, R., Michael, A., 2012. Generation and characterization of chicken egg yolk antibodies against Propionibacterium acnes for the prevention of acne vulgaris. Indian Journal of Dermatology. 57, 1519.

Sharma, J.M., 1997. The structure and function of the avian immune system. Acta Vet Hung 45, 229,238.

Shen, H., Cai, Y., Zhang, H., Wu, J., Ye, L., Yang, P., Lin, X., Jiang, S., Liao, M., 2021. Anti-SARS-CoV-2 Igy Isolated From Egg Yolks of Hens Immunized With Inactivated Sars-Cov-2 for Immunoprophylaxis of COVID-19. Virol Sin. 36, 1080–1082.

Shen, Y., Ziquan, Lv., Yang, L., Liu, D., Yanran, Ou., Chunyan, Xu., Weiwen, Liu., Dongmei, Yuan., Yuxin Hao, Junjia He, Xing Li, Yuqing Zhou, Timothy R. Walsh, Jianzhong Shen, Junjie Xia, Yuebin Ke, Yang Wang., 2019. Integrated aquaculture contributes to the transfer of mcr-1 between animals and humans via the aquaculture supply chain, Environment International. 130, 104708.

Shimizu, M., Nagashima, H., Sano, K., Hashimoto, K., Ozeki, M., Tsuda, K., 1992. Molecular stability of chicken and rabbit immunoglobulin G. Biosci Biotechnol Biochem. 56, 270–274.

Shimizu, M., Nakai, S., Fitzsimmons, R.C., 1988. An-*E. coli* immunoglobulin Y isolatedfromeggyolk of immunizedchickens as a potentialfoodingredient. J. Food Sci., 5, 1360–1366.

Skrabanja, A.T., De, Meere A.L., De Ruiter, R.A., Van Den Oetelaar P.J., 1994. Lyophilization of biotechnology products. PDA. J. Pharm. Sci. Technol. 48(6), 311317.

Spillner, E., Braren, I., Greunke, K., Seismann, H., Blank, S., du, Plessis, D., 2012. Avian IgY antibodies and their recombinant equivalents in research, diagnostics and therapy. Biologicals. 40(5), 313–322.

Staak, C., Schwarzkopf, C., Behn, I., Hommel, U., Hlinak, A., Schade, R., Erhard, M., Isolation of IgY from Yolk In: Schade, R., Behn, I., Erhard, M., Hlinak, A., Staak, C., 2001. Editors Chicken egg yolk antibodies, production and application: IgY-technology. Berlin (Heidelberg): Springer Berlin Heidelberg. 2001. p. 65–107.

Taha, M., Almeida, M.R., Silva, F.A., Domingues, P., Ventura, S.P., Coutinho, J.A., Freire, M.G., 2015. Novel biocompatible and selfbuffering ionic liquids for biopharmaceutical applications. Chemistry (Weinheim an der Bergstrasse, Germany). 21,47814788.

Thu, H.M., Myat T.W., Win, M.M., Thant, K.Z., Rahman, S., Umeda, K., Nguyen, S.V., Icatlo, F.C Jr., Higo-Moriguchi, K., Taniguchi, K., Tsuji, T., Oguma, K., Kim, S.J., Bae, H.S., Choi, H.J., 2017. Chicken Egg Yolk Antibodies (IgY) for Prophylaxis and Treatment of Rotavirus Diarrhea in Human and Animal Neonates: A Concise Review. Korean J Food Sci Anim Resour. 37(1), 1-9.

Tsubokura, K., Berndtson, E., Bogstedt, A., Kaijser, B., Kim, M., 1997. Oral administration of antibodies as prophylaxis and therapy in *Campylobacter jejuni* infected chickens. Clin. Immunol. 108: 451–455.

van Regenmortel, M.H.V., 1993. Eggs as protein and antibody factories. In Proceedings of the European Symposium on the Quality of Poultry Meat. 257–263.

Wang, L.H., Li, X.Y., Jin, L.J., You, J.S., Zhou, Y., Li, S.Y. and Xu, Y.P., 2011. Characterization of chicken egg yolk immunoglobulins (IgYs) specific for the most prevalent capsular serotypes of mastitis-causing *Staphylococcus aureus*. Vet. Microbiol. 149, 415-421.

Warr, G.W., Magor, K.E., Higgins, D.A., 1995. Igy: Clues to the Origins of Modern Antibodies. Immunol Today. 16,392–398.

Wei, S., Duan, S., Liu, X., Wang, H., Ding, S., Chen, Y., Xie, J., Tian, J., Yu, N., Ge, P., Zhang, X., Chen, X., Li, Y., Meng, Q., 2021. Chicken Egg Yolk Antibodies (Igys) Block the Binding of Multiple Sars-Cov-2 Spike Protein Variants to Human Ace2. Int Immunopharmacol. 90, 107172.

Woolley, J.A., Landon, J., 1995. Comparison of antibody production to human interleukin-6 (IL-6) by sheep and chickens. J Immunol Methods. 178,253–65.

Wu, C.J., Wang, H., Chan, Y.L., Li, T.L., 2011. Passive immune-protection of small abalone against *Vibrio alginolyticus* infection by anti-Vibrio IgY-encapsulated feed. Fish and Shellfish Immunology 30, 1042–1048.

Xie, M.Q., Meng, Y.X., Li, Z.H., Li, Y.Q., Zhang K.X., Zhao X.M., Xiong G.J, Chen Y.M., 2004. Effect of specific immunoglobulin Y in the treatment of acute and chronic pharyngitis. Zhonghua Er Bi Yan Hou Ke Za Zhi; 39,112–115.

Xu, L., Jian Che., Yongping Xu., Yan Chen., Yuan Li., Bilal Murtaza., Lili Wang., Meixia Zhang., Xiaoyu Li., 2020. Oral administration of microencapsulated egg yolk immunoglobulin (IgY) in turbot (*Scophthalmus maximus*) to combat against Edwardsiella tarda 2CDM001 infections, Fish and Shellfish Immunology. 106, 609-620,

Xu, Y., Li, X., Jin, L., Zhen, Y., Lu, Y., Li, S., You, J., Wang, L., 2011. Application of chicken egg yolk immunoglobulins in the control of terrestrial and aquatic animal diseases: A review. Biotechnol Adv. 29(6), 860-8.

Yang, Y., Cui, L., Mikeremu, S., Zhang, X.Y., 2018. Generation and Characterization of Chicken Egg Yolk Antibodies against Bovine Interleukin-6. Applied Engineering in Agriculture. 34(5).

Yokoyama, H., Peralta, R.C., Sendo, S., Ikemori, Y., Kodama, Y., 1993. Detection of Passage and Absorption of Chicken Egg Yolk Immunoglobulins in the Gastrointestinal Tract of Pigs by Use of Enzyme-Linked Immunosorbent Assay and Fluorescent Antibody Testing. Am J Vet Res. 54, 867–72.

Young, C.A., Silversides, F.G., Jones, S.R.M., 2007. Chicken-derived IgY recognizes developing and mature stages of *Loma salmonae* (Microsporidia) in Pacific salmon, *Oncorhynchus* spp., Aquaculture. 273(4), 398-404.

Zhang, M., Huijun Geng., Muhammad Tariq Javed., Le Xu., Xiaoyu Li., Lili Wang., Shuying Li., Yongping Xu., 2021. Passive protection of Japanese pufferfish (*Takifugu rubripes*) against *Vibrio harveyi* infection using chicken egg yolk immunoglobulins (IgY), Aquaculture. 532.

Zhen, Y.H., Jin, L.J., Guo, J., Li, X.Y., Lu, Y.N., Chen, J. and Xu, Y.P., 2008. Characterization of specific egg yolk immunoglobulin (IgY) against mastitiscausing *Escherichia coli*. Vet. Microbiol. 130, 126-133.

Index

D

E

F

G

N

O